Butterflies
of BRITISH COLUMBIA

John Acorn
Ian Sheldon

Lone Pine Publishing

© 2006 by John Acorn, Ian Sheldon and Lone Pine Publishing
First printed in 2006   10 9 8 7 6 5 4 3 2 1
Printed in China

All rights reserved. No part of this work covered by the copyrights hereon may be reproduced or used in any form or by any means—graphic, electronic or mechanical—without the prior written permission of the publisher, except for reviewers, who may quote brief passages. Any request for photocopying, recording, taping or storage on information retrieval systems of any part of this work shall be directed in writing to the publisher.

The Publisher: Lone Pine Publishing

10145–81 Avenue
Edmonton, AB
T6E 1W9
Canada

1808 B Street NW, Suite 140
Auburn, WA, USA 98001

Website: www.lonepinepublishing.com

Library and Archives Canada Cataloguing in Publication

Acorn, John, 1958–
    Butterflies of British Columbia / John Acorn, Ian Sheldon.

Includes index.
ISBN-13: 978-1-55105-113-0
ISBN-10: 1-55105-113-3

    1. Butterflies—British Columbia—Identification.
I. Sheldon, Ian, 1971- II. Title.

QL552.A363 2006   595.78'909711   C2006-902742-0

Editorial Director: Nancy Foulds
Project Editor: Sandra Bit
Production Manager: Gene Longson
Book Design & Layout: Heather Markham
Production: Heather Markham, Curtis Pillipow, Willa Kung
Cover Design: Gerry Dotto
Image Scanning: Elite Lithographers Co., Ltd.

Photography: All photographs by John Acorn, with the exception of the photo of John Acorn on p. 15, taken by Marissa Shoemaker, the photo of John Acorn with Robert M. Pyle on p. 26, taken by Thea L. Pyle and the photo of Ian Sheldon and John Acorn on p. 360, taken by Dena Stockburger. All photos used with permission.
Illustrations: All illustrations by Ian Sheldon.
Cover illustration: Satyr Comma; *title page illustration:* Blue Copper

We acknowledge the financial support of the Government of Canada through the Book Publishing Industry Development Program (BPIDP) for our publishing activities.

PC: P13

TABLE OF CONTENTS

Dedications, p. 4
Acknowledgements, p. 5
Quick Reference Guide, p. 6
Introduction, p. 15

Swallowtails & Parnassians. . 32
   Parnassians 33
   Swallowtails. 40
   Old World
      Swallowtails. 42
   Tiger Swallowtails. 49

Whites & Sulphurs 58
   Whites, Marbles &
      Orangetips 60
   Sulphurs 87

Gossamer-Winged. 108
   Coppers 110
   Hairstreaks &
      Elfins 130
   Blues. 153

Metalmarks 184
Brush-Footed 188
   Fritillaries 190
   Crescents &
      Checkerspots. 231
   Nymphs 252
   Anglewings & Their
      Relatives 253
   Admirals 274
   Satyrs 277
   Alpines 285
   Arctics 296
   Milkweed Butterflies 307

Skippers 310
   Spread-winged 312
   Grass. 325

Glossary, p. 344
Resources, p. 347
Checklist, p. 348
Index of Scientific Names, p. 353
Index of Common Names, p. 356
About the Authors, p. 360

To Felix Sperling, a great scientist and a great friend.

John Acorn, author

To Sally Corbett

For entomological encouragement, support, musings and inspirations during my time at Cambridge.

Ian Sheldon, illustrator

Acknowledgements

A number of people were very helpful in the preparation of this book. Ian needed to borrow specimens for reference while painting, and these came from the Royal British Columbia Museum, the E.H. Strickland Entomological Museum (University of Alberta), and the collections of Chris Schmidt, Dave Lawrie, Terry Thormin and Gerry Hilchie. Thanks are also due to Rob Cannings, Felix Sperling and Danny Shpeley. Others, who contributed in important but perhaps more subtle ways, include Dick Beard, Cris Guppy, Mike Overton, Robert M. Pyle, Thomas Simonsen and Andy Warren. I would also like to thank both the Alberta Lepidopterists' Guild, and the Lepidopterists' Society, in a general sort of way, for helping make the study of butterflies both more rigorous and more enjoyable. At Lone Pine, Sandra Bit, Heather Markham, Curtis Pillipow, Nancy Foulds and Shane Kennedy did a superb job of bringing the book not only to completion but also to a level that exceeded our expectations. I would also like to thank my wife Dena, our sons Jesse and Benjamin, my sister Annalise, and my parents June and Glen Acorn, for their support and help along the way. This book was not peer-reviewed in the strict sense, and I freely admit that all opinions are my own. I hold my friends and colleagues blameless for these, and I look forward to many discussions in response to the things I have written here. Hopefully, the majority of them turn out to be more or less true.—*John Acorn*

6 QUICK REFERENCE GUIDE

Use this quick reference (arranged in the same order in which the butterflies appear in the book) as a guide to the butterfly groups, and as a reminder of the details once you are better acquainted with the species.

PARNASSIANS

Rocky Mountain Parnassian
45–70 mm • p. 34

Clodius Parnassian
53–82 mm • p. 36

Eversmann's Parnassian
53–61 mm • p. 38

Old World Swallowtail
60–100 mm • p. 44

Anise Swallowtail
60–95 mm • p. 46

Indra Swallowtail
65–85 mm • p. 48

Western Tiger Swallowtail
75–105 mm • p. 50

Canadian Tiger Swallowtail
70–105 mm • p. 52

WHITES AND SULPHURS

Pale Swallowtail
75–110 mm • p. 54

Two-tailed Swallowtail
85–120 mm • p. 56

Pine White
45–55 mm • p. 62

Western White
35–50 mm • p. 64

Checkered White
35–50 mm • p. 66

Becker's White
35–50 mm • p. 68

Spring White
35–45 mm • p. 70

Veined White
40–50 mm • p. 72

Margined White
40–50 mm • p. 73

Arctic White
35–45 mm • p. 74

Cabbage White
35–50 mm • p. 75

Large Marble
35–55 mm • p. 77

QUICK REFERENCE GUIDE 7

WHITES AND SULPHURS

Northern Marble
35–45 mm • p. 79

Green Marble
35–40 mm • p. 81

Desert Marble
40–45 mm • p. 82

Sara Orangetip
40–45 mm • p. 83

Stella Orangetip
40–50 mm • p. 85

Clouded Sulphur
35–55 mm • p. 89

Orange Sulphur
35–60 mm • p. 91

Giant Sulphur
40–60 mm • p. 93

Pink-edged Sulphur
40–50 mm • p. 94

Pelidne Sulphur
35–45 mm • p. 95

Palaeno Sulphur
40–45 mm • p. 96

Western Sulphur
45–55 mm • p. 97

Alexandra's Sulphur
40–60 mm • p. 98

Christina's Sulphur
35–55 mm • p. 100

Mead's Sulphur
35–45 mm • p. 102

Hecla Sulphur
35–45 mm • p. 104

GOSSAMER-WINGED BUTTERFLIES

Canada Sulphur
35–50 mm • p. 105

Nastes Sulphur
30–45 mm • p. 106

American Copper
25–35 mm • p. 112

Lustrous Copper
25–30 mm • p. 114

8 QUICK REFERENCE GUIDE

GOSSAMER-WINGED BUTTERFLIES

Bronze Copper
35–40 mm • p. 116

Gray Copper
35–40 mm • p. 118

Blue Copper
30–35 mm • p. 120

Dorcas Copper
22–30 mm • p. 122

Purplish Copper
25–35 mm • p. 124

Lilac-bordered Copper
30–35 mm • p. 126

Mariposa Copper
25–30 mm • p. 128

Coral Hairstreak
30–35 mm • p. 132

Behr's Hairstreak
28–30 mm • p. 133

Sooty Hairstreak
25–30 mm • p. 134

California Hairstreak
30–35 mm • p. 135

Sylvan Hairstreak
25–35 mm • p. 136

Striped Hairstreak
25–30 mm • p. 137

Hedgerow Hairstreak
20–32 mm • p. 138

Bramble Hairstreak
22–30 mm • p. 139

Sheridan's Hairstreak
22–26 mm • p. 140

Thicket Hairstreak
25–30 mm • p. 141

Johnson's Hairstreak
30–35 mm • p. 142

Cedar Hairstreak
25–30 mm • p. 143

Juniper Hairstreak
25–30 mm • p. 144

QUICK REFERENCE GUIDE 9

GOSSAMER-WINGED BUTTERFLIES

Brown Elfin　　Western Elfin　　Moss's Elfin　　Hoary Elfin
20–28 mm • p. 145　28–30 mm • p. 147　25–30 mm • p. 148　22–30 mm • p. 149

Eastern Pine Elfin　Western Pine Elfin　Grey Hairstreak　Eastern Tailed Blue
25–30 mm • p.150　28–32 mm • p. 151　25–30 mm • p. 152　28–30 mm • p.154

Western Tailed Blue　Spring Azure　Western Spring Azure　Square-spotted Blue
28–30 mm • p. 156　28–30 mm • p. 158　28–30 mm • p. 160　20–28 mm • p. 162

Arrowhead Blue　Silvery Blue　Northern Blue　Anna's Blue　Melissa's Blue
27–35 mm • p. 164　22–30 mm • p. 166　20–30 mm • p. 168　30–32 mm • p. 170　20–30 mm • p. 172

Greenish Blue　Boisduval's Blue　Acmon Blue　Cranberry Blue　Arctic Blue
22–30 mm • p. 174　25–35 mm • p. 176　20–30 mm • p. 178　22–28 mm • p. 180　25–28 mm • p. 182

METALMARKS · BRUSH-FOOTED BUTTERFLIES

Mormon Metalmark
25-35 mm • p. 186

Variegated Fritillary
45-65 mm • p. 192

Great Spangled Fritillary
65-90 mm • p. 194

Callippe Fritillary
60-65 mm • p. 196

Zerene Fritillary
55-65 mm • p. 198

Aphrodite Fritillary
55-65 mm • p. 200

Atlantis Fritillary
55-65 mm • p. 202

Northwestern Fritillary
45-60 mm • p. 204

Hydaspe Fritillary
45-60 mm • p. 206

Mormon Fritillary
40-50 mm • p. 207

Mountain Fritillary
30-40 mm • p. 208

Bog Fritillary
40 mm • p. 210

Silver-bordered Fritillary
40-45 mm • p. 212

Meadow Fritillary
45 mm • p. 214

Frigga Fritillary
40 mm • p. 216

Dingy Fritillary
35 mm • p. 218

Western Meadow Fritillary
40 mm • p. 220

Polar Fritillary
40-45 mm • p. 221

Alberta Fritillary
45 mm • p. 222

Freija Fritillary
40 mm • p. 224

QUICK REFERENCE GUIDE 11

BRUSH-FOOTED BUTTERFLIES

Beringian Fritillary
45–50 mm • p. 225

Astarte Fritillary
45–50 mm • p. 226

Distinct Fritillary
45–50 mm • p. 228

Titania Fritillary
40–45 mm • p. 230

Northern Crescent
35–40 mm • p. 232

Tawny Crescent
32–40 • p. 234

Field Crescent
35–40 mm • p. 236

Pale Crescent
45 mm • p. 238

Mylitta Crescent
35 mm • p. 240

Northern Checkerspot
35–40 mm • p. 242

Rockslide Checkerspot
35–40 mm • p. 244

Hoffman's Checkerspot
35–40 mm • p. 245

Gillett's Checkerspot
35–45 mm • p. 246

Edith's Checkerspot
30–45 mm • p. 248

Variable Checkerspot
40–55 mm • p. 250

Satyr Comma
40–55 mm • p. 254

Green Comma
40–50 mm • p. 256

Hoary Comma
40–50 mm • p. 257

Oreas Comma
50 mm • p. 259

Gray Comma
50 mm • p. 260

BRUSH-FOOTED BUTTERFLIES

Compton Tortoiseshell
55–75 mm • p. 262

California Tortoiseshell
50–60 mm • p. 264

Mourning Cloak
50–80 mm • p. 266

Milbert's Tortoiseshell
40–55 mm • p. 268

Painted Lady
50–70 mm • p. 269

West Coast Lady
40–55 mm • p. 271

American Lady
40–60 mm • p. 272

Red Admirable
50–60 mm • p. 273

Lorquin's Admiral
50–70 mm • p. 275

White Admiral
50–70 mm • p. 276

Ringlet
30–40 mm • p. 278

Common Wood Nymph
50–60 mm • p. 279

Great Basin Wood Nymph
40–45 mm • p. 281

Dark Wood Nymph
40–45 mm • p. 283

Common Alpine
35–45 mm • p. 287

Taiga Alpine
35–45 mm • p. 289

Vidler's Alpine
35–45 mm • p. 290

Ross's Alpine
35–45 mm • p. 291

Red-disked Alpine
35–45 mm • p. 292

Magdalena Alpine
44–50 mm • p. 293

QUICK REFERENCE GUIDE 13

BRUSH-FOOTED BUTTERFLIES

Mt. McKinley Alpine
45–50 mm • p. 294

Mountain Alpine
30–40 mm • p. 295

Great Arctic
55–65 mm • p. 297

Macoun's Arctic
55–65 • p. 298

Chryxus Arctic
40–55 mm • p. 299

Uhler's Arctic
35–45 mm • p. 300

Alberta Arctic
35–45 mm • p. 301

Jutta Arctic
35–55 mm • p. 302

White-veined Arctic
40–50 mm • p. 303

Melissa Arctic
35–50 mm • p. 304

Polyxenes Arctic
35–50 mm • p. 305

Philip's Arctic
45–55 mm • p. 306

SKIPPERS

Monarch
90–100 mm • p. 308

Silver-spotted Skipper
45–50 mm • p. 313

Northern Cloudywing
35–45 mm • p. 314

Dreamy Duskywing
25–35 mm • p. 315

Propertius Duskywing
35–45 mm • p. 316

Pacuvius Duskywing
35–40 mm • p. 317

Afranius Duskywing
30–35 mm • p. 318

Persius Duskywing
27–35 mm • p. 319

SKIPPERS

Grizzled Skipper 25–30 mm • p. 320 Two-banded Checkered Skipper 23–28 mm • p. 321 Common Checkered Skipper 25–32 mm • p. 322 Common Sootywing 25–30 mm • p. 324

Arctic Skipper 20–30 mm • p. 326 Garita Skipper 20–25 mm • p. 327 European Skipper 20–25 mm • p. 328 Common Branded Skipper 25–30 mm • p. 329 Plains Skipper 20–30 mm • p. 330

Western Branded Skipper 25–30 mm • p. 331 Juba Skipper 30–35 mm • p. 332 Nevada Skipper 25–30 mm • p. 333 Sachem Skipper 25–30 mm • p. 334

Peck's Skipper 20–30 mm • p. 335 Sandhill Skipper 20–25 mm • p. 336 Draco Skipper 20–30 mm • p. 337 Tawny-edged Skipper 20–30 mm • p. 338 Long Dash 25–30 mm • p. 339

Sonoran Skipper 25–30 mm • p. 340 Woodland Skipper 25–30 mm • p. 341 Dun Skipper 25–30 mm • p. 342 Common Roadside Skipper 20–25 mm • p. 343

Introduction

John Acorn   *Ian Sheldon*

Almost 100 years ago, if you were interested in the butterflies of British Columbia, perhaps you might have purchased a copy of Dr. Clarence Weed's book, *Canadian Butterflies Worth Knowing* (1923). Inside, you would have been surprised to find a text written almost entirely about the butterflies of New England and points south. In fact, the exact same book had already been published in 1917 under the title *Butterflies Worth Knowing*, and was clearly written for an audience of Americans, not Canadians, and easterners, not folks from British Columbia. Thus, a century ago it was clear to some, at least, that the butterflies of British Columbia did not qualify as "worth knowing," and were seen as only trivially Canadian.

At the time, the only butterfly book to cover the entirety of the United States and Canada was W.J. Holland's 1898 volume, *The Butterfly Book*, a book that was only marginally suited to the identification of our northwestern fauna. British Columbians had to wait until 1961 for the first relatively modern treatment of the North American butterflies (Ehrlich and Ehrlich's black and white, coil-bound *How to Know the Butterflies*).

They then had to wait another 15 years until the first field guide to western butterflies was published, in the Peterson Field Guide series, authored by J. W. Tilden and Arthur C. Smith. To the south, a long tradition of butterfly guides had developed in California, but during the time of my childhood (and I am not an old man), British Columbia was still in its pioneer period with respect to the appreciation and study of butterflies.

When Ian Sheldon and I began working on this book, there were two prominent western butterfly guides (Peterson's *A Field Guide to Western Butterflies* [1999] and the Oxford University Press *Butterflies Through Binoculars* series), and two very fine North American guides in print (one by James A. Scott, *Butterflies of North America* [1986], and one by Robert M. Pyle, *The Audubon Society Field Guide to North American Butterflies* [1981]), along with a book called the *Butterflies of Canada* (1998), by Ross Layberry, Peter Hall and Don Lafontaine. More recently, two other books have appeared on the butterflies of this region. Most notably, Jon H. Shepard and Crispin S. Guppy have written a comprehensive treatment of the British Columbian butterfly fauna entitled *Butterflies of British Columbia*, published in 2001. As well, biologist Robert M. Pyle has written a lovely book entitled *Butterflies of Cascadia* (2002), with Cascadia being a region that includes much of British Columbia's southern Interior. If you find yourself truly caught up in the love of butterflies, you will eventually want to own them all.

Why did Ian and I produce yet another book? Well, for one thing, we believed there was room for a smaller, less technical book about British Columbia butterflies. I also see the book as a showcase for Ian's fine illustrations, with the primary purpose of my text being to provide a sort of extended commentary on the paintings. Ian has depicted each species in its most typical natural pose, and for those that rarely spread their wings when at rest, the butterfly is painted as if in flight.

We both agree with Vladimir Nabokov, the famous lepidopterist and novelist, who once wrote, "I personally belong to a category of curieux who, in order to aquaint themselves properly with a butterfly and to visualize it, require three things; its artistic depiction, a compendium of all that has been written about it, and its insertion within the general system of classification." (*Father's Butterflies*, reprinted in the *Atlantic Monthly*, April 2000.)

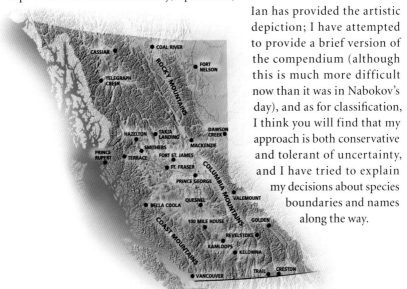

Ian has provided the artistic depiction; I have attempted to provide a brief version of the compendium (although this is much more difficult now than it was in Nabokov's day), and as for classification, I think you will find that my approach is both conservative and tolerant of uncertainty, and I have tried to explain my decisions about species boundaries and names along the way.

Whites as a group are taxonomically challenging.

Butterfly Names

Everyone who studies butterflies must somehow make peace with their names. You might think this would be a simple matter, but in fact, it is not. Why? Well, to begin, some people strongly prefer English to scientific names, and others prefer the opposite. In a perfect world the two sets of names would be synonyms. Scientific species names are two-part names, such as *Speyeria atlantis,* and do not always reflect the English name. The first word is the genus, or group name, and the two words together form the species name. The second word alone is called the "specific epithet." You might think that the names of such well-known creatures as butterflies would be stable by now, but while the never-ending ritual of the English *versus* scientific names debate rages on, the names themselves change at a rate that shows no signs of slowing, let alone coming to a halt.

I personally have learned to take some pleasure from the names themselves, and I highly recommend that you learn both the scientific and the English names, as well as the subspecies names if you have the energy. This will serve you well as names shift and shuffle, appear and vanish. Subspecies, or geographic races (three-part scientific names, such as *Speyeria atlantis hollandi*), are often "elevated" to species status (although the opposite happens as well), and subspecies are also often easier to recognize in the field than the species as a whole, since the species as a whole may be confusingly variable across its entire geographic range. I also suggest that you learn the taxonomic groupings of butterflies, which is why I have provided summaries at the beginning of each such group. Families are divided into subfamilies, subfamilies are divided into tribes, tribes are divided into genera, genera are divided into subgenera, or species groups, these latter categories are divided into species, and species are divided into subspecies or "geographic races." The most important rank is that of species. Species are represented as

Veined White, Pieris oleracea

the "branches" of the evolutionary tree of life, which is why the butterfly accounts that follow describe each species separately.

Scientific names, which are formulated according to the rules of Latin grammar (but are not necessarily Latin in origin, and should not be called "Latin names"), have the advantage that they allow communication with a broad, international group of butterfolk. English names have the advantage of familiarity and ease of pronunciation. There is, by the way, no standard for the pronunciation of scientific names, although there are a number of systems for pronunciation of Latin, none of which is officially endorsed by the science of biology.

Why do butterfly names change? The main factors that contribute to this phenomenon are as follows. 1) Different people have different ideas of what constitutes a species. 2) New ecological information allows more accurate recognition of the boundaries between species. 3) New analyses of relationships among butterflies allow researchers to group them in more meaningful ways. 4) Some people ("lumpers") seem to prefer larger groups and fewer names, emphasizing similarities. 5) Some people ("splitters") prefer smaller groups and more names, emphasizing differences. 6) Scientific names are governed by formal rules that require correction if the rules are broken, and thus some old names need to be replaced. And 7) there is no truly official system for English names and, thus, not all authors share the same views as to which name is correct.

Perhaps no other group demonstrates those reasons as well as the Mustard Whites. Let's take a few pages to review this story, in the hopes that it will, at the very least, demystify the perpetual indecision that typifies this field.

For decades, the Mustard White (*Pieris napi*) was considered one species in North America; the same widespread butterfly that Europeans generally call the Green-veined White. But those with a broad knowledge of these butterflies knew the situation was not clear-cut. In W.J. Holland's words, "this is a Protean species, of which there exists many forms." Likewise, John H. and Anna Comstock remarked in *How to Know the Butterflies* (1904), "this innocent white butterfly is a source of dire confusion, because its history is so intricate and it masquerades in so many guises." Truth is, they didn't know the half of it. This is a clear instance where the noise of individual variation is about as strong as the signal that distinguishes one species from another. It really is "a mess of overwhelming proportions," as lepidopterist Art Shapiro once wrote.

It was reasonable, then, that the species should be subdivided. A lepidopterist named B.C.S. Warren tried in 1968, and came to the conclusion that there are five separate species of Mustard Whites in North America. However, his classification was not widely adopted, in part because it was based almost entirely on the form of the "sex scales" on the male wings. Then came Ulf Eitschberger, a pharmacist and amateur lepidopterist in Germany, who produced what I have heard some taxonomists call "a major brain dump." Almost no one I know owns a copy, in part because it is written in German, fills two volumes, runs 1005 pages, contains hundreds of plates depicting thousands of specimens, and sells for about $500. It is a huge effort, but a labour of love. It was also badly timed, in retrospect, because 1983 was not a good time for Eitschberger to be messing with butterfly taxonomy from an amateur position.

The Cabbage White, a white butterfly with an uncontroversial name

New techniques were revolutionizing the field. In particular, enzyme electrophoresis (the study of different forms of the same enzymes, and how they correspond to different forms of the same genes) seemed like the best means of determining whether populations did or did not interbreed, and therefore whether they were part of one species or more. As well, the science that had been known as "taxonomy" had recently been renamed "systematics," and was now based on a more advanced set of theories and analytical techniques, unfamiliar to those outside inner scientific circles. Eitschberger was a master of none of these things: he was an old-fashioned butterfly collector with aspirations of solving a long-standing problem in butterfly science.

Eitschberger's book was reviewed in a diplomatic but unfavourable light by the top American expert on the subject, Arthur Shapiro, a prominent lepidopterist at the University of California at Davis. Shapiro, who is the namesake of one of Eitschberger's new subspecies of white, was firm but gentle in his criticisms, using such terms as "unsatisfactory" and "not at all like what one has come to hope for in these sophisticated times." It was clear that the era of the amateur butterfly taxonomist had come to a close, and that although Shapiro had sympathy for his German colleague, he was not about to forgive his naïveté with regards to matters of modern biology.

Another of Eitschberger's critics was Otakar Kudrna, whose doctoral thesis, written in Switzerland, amounted to a thundering critique of Eitschberger's work. Kudrna had muddied the waters in another way in the decade preceding 1983, suggesting that the mustard whites belonged in the genus *Artogeia*, not *Pieris*, and then reversing his own opinion once a few major field guides had adopted this system. Then, he and another Mustard White specialist, Hansjürg Geiger, published a shortened version of his thesis, and suggested that the new species names proposed in Eitschberger's book should be placed on the Official Index of Rejected and Invalid Specific Names in Zoology. (That is a bad thing when it happens—really bad—but it didn't come to pass.) Kudrna and Geiger were merciless in their critique of Eitschberger's opus, writing that "it is not only without value, it is irresponsible." They went on to say that his new species were "based on complete intuition," and that "we deeply regret that it was ever published, because it brings the science of taxonomy and lepidopterology into disrepute."

Who knows what poor Ulf Eitschberger thought of all this. Art Shapiro confided in his review, "Eitschberger told me over a stein of beer that he hopes other people will take up and expand his work. That is good, for it must—and will—be done." Sure enough, in 1991, Geiger and Shapiro got together and published their own classification of the Mustard White group, based mostly on enzyme electrophoresis. It bore a reasonably close resemblance to both Warren's and Eitschberger's views, at least in their general outlines, but was much better received, no doubt, because people were

Sulphurs are confusing too, but in the hand it is clear that this is a Clouded Sulphur.

getting used to the idea that multiple species of mustard whites really existed.

Today, although butterfly experts quibble over details, they generally recognize about a half dozen species of Mustard Whites on this side of the Bering Sea, with three in British Columbia, along with a close relative, the familiar Cabbage White. Still, however, there are some who insist that there are only two Mustard Whites in North America: the Mustard White in the broad sense (*P. napi*), and the West Virginia White (*P. virginiensis*). In retrospect, we can see that this story is the product of a confusing group of butterflies, the desire of an amateur to make his mark on the science that he loved and the desire of professionals to keep their turf to themselves. It is no wonder people unfamiliar with the inner workings of butterfly science find the whole thing confusing and frustrating.

Slowly, I have come to the realization that the main reason our butterfly names are not stable is that we are not yet finished the job of fully studying and understanding our butterflies. So, as much as I would love to be able to tell you that this book will be timeless, I know for a fact that the next butterfly guide to come into print after this one will use different names for at least some of the species. If you are ready for this, you will not dismiss the following perfectly good information on the mistaken belief that it is outdated. I use not one, but a whole cabinet full of butterfly books, and I wouldn't part with a single volume, especially for a silly reason such as its age!

Purplish Copper eggs

Alberta Arctic egg

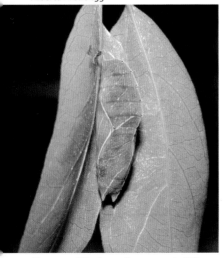
The pupa of the Tawny Emperor, not a B.C. species

Butterfly Biology

Before you get the mistaken impression that the study of butterflies is just a study of words, we should also quickly review a few things about the lives of butterflies. Butterflies begin life as an egg, and mother butterflies typically lay hundreds of eggs on the appropriate food plant for the caterpillars, and do not look after the eggs or the young. Young butterflies are called caterpillars, or more technically larvae (the plural of larva). All of the caterpillars of British Columbian butterflies eat leaves, seeds or flower parts, and they all shed their skins four times to become larger caterpillars. The periods between sheddings of the skin are called instars, or stadia. The word "instar" does not refer to the period between moults, as most people use it. For that, the correct word is stadium (plural: stadia). Instars refer to the periods between the internal development of the new cuticle, and they, therefore, begin before the moult, and the onset of the next stadium.

There is no simple way to distinguish the caterpillars of butterflies from those of moths, although familiarity with the various families of butterfly caterpillars will help. For this reason, and because most caterpillars change in appearance as they grow, or appear in two or more different colour morphs, Ian Sheldon and I decided not to include caterpillars in this book.

The final shedding of the caterpillar skin reveals a resting stage called the pupa, or chrysalis. If the pupa is enclosed by silk, the silk is called a cocoon, but most butterflies, unlike

Silver-spotted Skipper caterpillar, out of its leaf nest, feeding.

most moths, do not spin a coccoon. Some butterfly caterpillars spin a thin silk belt or "girdle" to support the pupa, however. The pupa remains more or less motionless as the body of the caterpillar transforms into the body of the adult butterfly, and then splits to allow the adult to emerge. This entire process is called "complete metamorphosis" and it is common to butterflies, moths and a variety of other insects including beetles, bees, ants, wasps, caddisflies and others.

Butterflies are a subgroup within the moths, best recognized by their "clubbed" antennae with thickened tips. Butterflies and moths together form the Lepidoptera, named for their scaly wings. The colour patterns on the wings of lepidopterans come from tiny, overlapping scales. The scales are modified hairs, and the closest relatives of the Lepidoptera, the caddisflies (or Trichoptera), have hairy wings instead of scaly wings, showing us what the ancient ancestor of the first moths might have looked like.

Butterflies are generally active when it is sunny. One fellow butterfly fan likes to say that butterflies are active when you can comfortably go outdoors in a t-shirt. Each species has its own characteristic flight season, however, so you will find some in spring, some in summer, and some all year round. For the most part, individual butterflies live only a week or so, but the combined flight season for a species may extend over a month or more each year. Some butterflies manage many generations per year, but for most there is only one. Some have partial second broods late in the season, the result of caterpillars that got a good head start on their buddies and were able to complete their metamorphosis before the end of the butterfly season. In British Columbia, although it is theoretically possible to find a butterfly on any day of the year, the season generally runs from March through October.

In some butterfly species, the male and female are differently patterned,

The Study of Butterflies

but in all butterflies the female is slightly larger and heavier because she carries the eggs. In old books, the male and female were called the "cock" and "hen," and although I'd love to push for a return to this quaint usage (along with "primaries" and "secondaries" for the front and hind wings), I think it is best to go along with the current practice of referring to them simply as male and female. For male butterflies, life is mostly about finding females, and to a lesser extent about finding food. In some species, the males patrol in search of females, and are on the wing for most of the day. These are generally our most familiar butterflies. In other species, the males choose a lookout perch from which they fly out to investigate potential mates. This habit makes these butterflies less obvious to the beginner. Female butterflies, on the other hand, spend most of their time searching for the appropriate host plant on which to lay their eggs.

To identify a butterfly, you generally have to look at its wings. The body, which is typically covered in hair, is rarely of interest to the field butterfly watcher, although there are exceptions. There are four wings, and of course, each wing has both an upper and a lower surface. The first pair of wings is the forewings, or front wings, or mesothoracic wings, or (in older books) primaries. The back pair of wings is called the hind wings, metathoracic wings, or secondaries. The wing surface that faces down when a butterfly spreads its wings is called the underwing, or bottom surface. The surface that faces up is called the upper or topside surface. It may also be called the inner surface because many butterflies fold their wings over their backs when at rest, hiding these surfaces from view. In this book, wingspan is given as the maximum span of a living butterfly with its wings spread, not the wingspread of

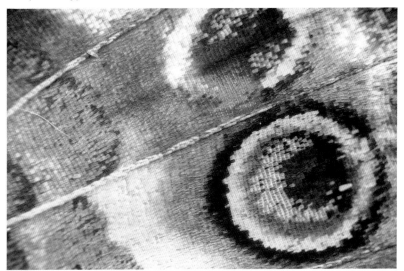

The scales on the wing of a Painted Lady produce a mosaic of colours and a complex pattern.

a pinned museum specimen. In a living butterfly, the leading edges of the forewings are generally held in a straight line when the butterfly is in flight. In a museum specimen, the trailing edges (not the leading edges) of the forewing are arranged in a straight line, perpendicular to the body.

On a butterfly wing, I refer to the anterior margin as the leading edge, and the posterior margin as the trailing or hind edge. The portion of the wing closest to the body is called the base, while the lateral margin of the wing is called the outer edge or outer margin. Many books call this the marginal region, and the area just inside the marginal region is called the submarginal region. I avoid these terms, because they often confused me as a youngster. Roughly at the centre of the wing there is an area without veins called the cell that is usually somewhat oval in shape. At the distal end (distal is the opposite of basal and refers to things that are far from the centre of the body) of the cell, there is often a cell-end spot or bar. For the most part, I have attempted to use plain English to describe the markings on butterfly wings.

Once you get the hang of butterfly identification, you will find yourself flipping quickly to the relevant part of the book when confronted with an unidentified butterfly. In the mean time, begin by scanning the thumbnail images in the quick-reference key at the front of the book to get an idea of the group in which the butterfly belongs. Read the identification tips carefully and fully, and be sure to compare what you see with the illustrations. I suspect that with practice, you will be able to identify almost all the butterflies you encounter.

I leave it to you whether you want to pursue butterflies with binoculars, with a camera or with a net and a box of collecting envelopes. I personally

Butterfly Wing Features

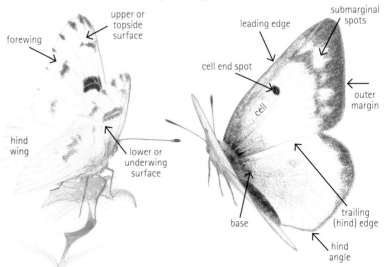

Robert M. Pyle, a superb writer, lecturer and butterfly conservationist from Washington State, with John Acorn

Lepidopterist Marie Djernaes explores a promising alpine meadow.

enjoy all three approaches. In my experience, there are things that are best learned through working with pinned specimens, but I realize that many people are not comfortable with this approach (or do not have the time and energy to put together a collection). If you are not a collector yourself, you may still find it valuable to examine a local museum collection to discover what the fuss is about. Butterfly collecting is a rare hobby that does plenty of good for the study of butterflies—don't get caught up in the current trend of blaming collectors for the decline of butterfly numbers. There is no proof of this, and collectors are generally a responsible lot. The Lepidopterists' Society actually publishes a set of collecting guidelines, if you are interested. The real evils in the world of butterflies are development, habitat loss and pesticides, and we should all be clear on this issue. Collectors are our friends, not foes. You will also find that a large collection of butterfly books is as useful as a large collection of butterfly specimens. I strongly advise that you get as many books as you can afford, and include some old ones in your library as well as new (after all, except for some shifts in geographic ranges, the butterflies themselves have not changed).

There are also advantages to photography and identifying butterflies from your photos. Digital cameras and email have made quick confirmation of field identifications easier than ever before, and a clear photo can often give you all the features you need to make a positive ID. Of course, if you really love butterflies, you will want to spend time with them in the field, and for that there is nothing like a good pair of close-focusing binoculars. I have used my Bausch and Lomb Elite 10X42 binoculars for years in this capacity (since they are also great for birding), but there are many good models on the market, and the close-focusing Elites are no longer in production. The Eagle Optics Platinum Class Rangers are another fine choice. As well, an exciting new development by Pentax (their inexpensive "Papilio" 6.5X21 model) has produced the first reverse-porro prism, close-focusing binocular with front lenses that are closer together. The alternative, close-focusing roof-prism binoculars, does not allow the butterfly to be seen with both eyes at once, at their closest focus. I would also promote the use of close-focusing monoculars in the field. I would expand on the use of this equipment, but experience has taught me that few people care, and most prefer to find out through their own experience which viewing tools they prefer.

If you want to contribute to our understanding of butterflies, my main recommendation is to keep field notes. For each butterfly you identify, try to keep a record of at least the location where you saw it and when. Some people define ecology as the

Let's face it—kids love to catch butterflies with nets!

study of distribution and abundance of living things, and so the distribution of butterflies in time and space is basic information. It is also the information on the labels that accompany butterfly specimens.

Incidentally, some of the most important markings on a butterfly are visible only in the ultraviolet (UV) range. Humans do not see UV light with the naked eye, but luckily there is an easy way to check for

Ultraviolet view of a pinned Sara Orangetip

Visible light view of the same butterfly

Prominent field guide author Paul Opler takes a macro photograph of a mud-puddling butterfly.

these reflections. Most video cameras and digital cameras are sensitive to UV, and there is a filter available from specialized camera stores (the Tiffen 18-A) that allows only UV, and no visible light, to pass through it. With this filter (and possibly a close-up lens as well) on your camera, you can check for UV reflections on male sulphurs. With free-flying sulphurs in the field, it is next to impossible. With pinned specimens, a circular "black light" bulb will give you the right sort of even illumination to check for reflections. With a live butterfly in the hand, you will need to carefully spread the wings with forceps while someone else looks through the camera. Make sure to examine the butterfly from a variety of angles; the reflections depend on such things. And remember— what you see through the camera is not exactly what the butterflies see. They see a mix of visible and UV reflections in a way that we can barely imagine.

Assessing butterfly abundance is much trickier than assessing distributions in time and space. Single-day butterfly counts, such as the Canada Day or Fourth of July counts are fun, but they are not as good for determining which butterflies are most common, or most rare, as are Christmas bird counts (on which the butterfly counts were based). To really get a good set of data on the butterflies of an area, I suggest going out in the same area at least once a week for the entire butterfly season, and counting all the butterflies you can identify for at least an hour each time. This is called a Pollard Walk, and data of this type can be extremely useful, sometimes decades after it was collected.

Rearing caterpillars is also a rewarding way to get to know butterflies. You may learn new things by demonstrating that a species can be reared on a plant that is not known to be a food plant. Even better, if you can find caterpillars in the field and make notes on their food plants, you can add to our knowledge in this way. Some of our butterflies are still unknown as caterpillars. I suppose it is possible to discover new species of butterflies even in British Columbia, but in the absence of such good fortune, at least there are mystery caterpillars to search for.

The study of butterflies is at an exciting point in its history. Large numbers of naturalists have begun to identify butterflies in the field, and these people have substantially increased the number of lepidopterists from the previous tiny band of collectors, who were the sole butterfly enthusiasts for many decades. The butterfly watching, or "butterflying," movement owes its origins to the works of Robert M. Pyle, a brilliant and talented scientist and author who also founded the Xerces Society for Invertebrate Conservation. You will find, however, that many newcomers to the butterfly world have taken an "out with the old and in with the new" attitude, and that everything from collecting, to scientific names, to professional lepidopterists in general have come under attack as old fashioned or out of touch. Please take these things with a grain of salt when you encounter them. Robert M. Pyle and I, along with other leaders in the butterfly world such as Jim Brock, Kenn Kaufman and Paul Opler, have all had opportunities to address the growing butterfly-watching movement, and I'm glad to report that we all share the same message—don't demonize the past, learn from it! The study of butterflies has some wonderful traditions and has been blessed with some brilliant scientists and authors. So, whether you carry a net, a camera, binoculars or just a walking stick and a curious mind, I hope you will approach the study of butterflies with warmth, humility and tolerance for others.

Traditional specimen photograph, showing variation among Christina's Sulphurs.

Illustrator's Note

As you flick through this field guide you may be given to wonder how it is that one butterfly has one illustration, and another has three or four. When John and I settled on the concept of this book, we wanted to create a field guide that reflected on the behavioural nature of the butterflies. If you are familiar with other field guides, you may have noticed that they used flat specimens from museum cases, or photos so small and lacking in detail that you couldn't always be sure about precise field marks (never mind field guides with tiny illustrations). We felt the best approach was to generously illustrate the entire book with butterflies in natural poses.

Different families of butterflies have a tendency to behave in different ways, from the egg right up to the flying adult. For example, the delightful adult hairstreaks flit quite happily about, but when they land they rest with their wings closed. It is a rare moment indeed that one gets to see the "inside," or more appropriately, the upper surface, of a living hairstreak. Thus we elected to illustrate only the lower surface of the butterfly, at rest. Other butterflies may have less predictable behavioural patterns. Take one of the blues for example (in the same family as the hairstreaks). The male blue is often dazzling in iridescent blues, the female is mostly brown, and the lower of both sexes is greyish with spots. Now the blues will often alight with wings partially spread as if to sun themselves, showing off their brown or blue colour, or they may close their wings to reveal the greys and the spots. We therefore felt the blues warranted three illustrations, one for the male topside, one for the female topside, and one for the underside markings (similar in both sexes). Unlike the hairstreaks, all of these postures and colours are readily viewable in nature.

The production department at Lone Pine Publishing has done a stellar job of presenting the illustrations. They've given an unusually generous amount of room to the illustrations, so please don't expect the Spring Azure to be as large as it is in the book; we'd be truly blessed if it was that big! Most of the butterflies are presented larger than life. Where there are multiple illustrations for a species, the first illustration in the layout is given the most room, and then the others (perhaps a male upper view and a lower view, for example) are presented slightly smaller on a facing page. This does not mean that the male is smaller, nor that the butterfly apparently shrinks when its wings are folded. Your best bet for size referencing is to pay attention to the wingspan notes in the text. If you are a newcomer to the world of butterflies, it will be a matter of days before you get your sense of proportions in tune with the dazzling world of butterflies.

While working on this book, many people asked where I got the photos in order to create such detailed illustrations. Well, I didn't. As someone attune to the intricacies of wing patterns and detail, I am often frustrated by the lack of information in photos. It is really hard to capture the true butterfly on film, "true" meaning with all the information. For those of you familiar with macro photography, you'll know that trying to get a good depth of field on a butterfly so small and with so many angles is a huge challenge. And let's face it, finding someone diligent enough to camp out on a mountain in northern British Columbia for weeks on end hoping to capture a close-up photograph, with good depth of field, of an obscure Satyr would be an impossibility! As a result of these complications, every single illustration has been done from a museum specimen. These almost always come flat, or "spread," but I have illustrated them in their natural poses. John and I have searched far and wide through private and public collections for such specimens. Thanks to the generous loan of countless museums and private collectors, I have been able to borrow and then illustrate all of these butterflies. We are indebted to these dedicated institutions and passionate individuals. Without them, this project would have been next to impossible.

The Butterflies

Swallowtails and Parnassians
(Family Papilionidae)

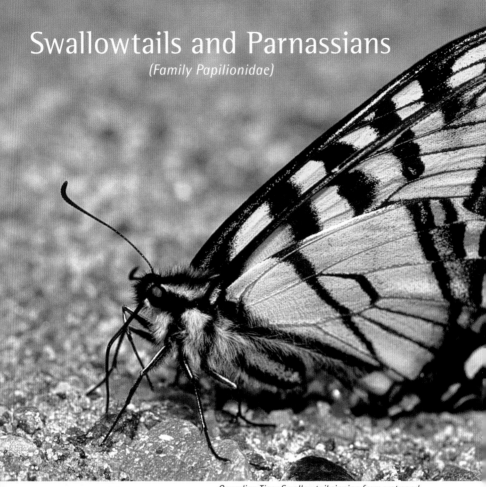

Canadian Tiger Swallowtail sipping from wet sand.

We begin our treatment of the British Columbia butterflies with the swallowtails and parnassians. Traditionally, these butterflies are treated first among the butterflies proper, but you will find that some books begin with the skippers instead. The skippers and the butterflies proper form the two main trunks of the evolutionary tree of butterflies, and each of the two groups has developed its own special features since they last shared a common ancestor. For that reason, I find it difficult to think of skippers as more primitive than the rest of the butterflies, but some authors are convinced that skippers are more moth-like and that true butterflies are more advanced. This way of thinking can be considered an echo of the outdated notion of the Great Chain of Being, on which all creatures can be arranged from simplest and primitive to most complex and advanced.

Modern biologists no longer think this way, and see all living creatures as equally successful and advanced in their own ways.

This, fortunately, allows us to begin with the butterflies that are most people's favourites. The swallowtail and parnassian family includes our largest and showiest species, as well as the largest butterflies in the world—the birdwings of Papua New Guinea. Get to know this family first, and use the enthusiasm they generate to fuel your study of the smaller, less colourful butterflies.

Parnassians
(Subfamily Parnassianiinae)

To find the classic "snow butterflies" of the mountains, you have to make a pilgrimage, and this pilgrimage is sure to be a pleasant one. Parnassians fly only in mid-summer when the alpine meadows are filled with flowers, and only on warm, sunny days when the weather is agreeable to us all. The group was named for Mt. Parnassus in Greece, which was sacred to the god Apollo, and indeed some people today still refer to these butterflies as "Apollos." One species in Europe is officially "The Apollo" (*Parnassius apollo*).

Parnassians are related to swallowtails, and in Asia you can find almost perfect intermediates between the two. Unlike swallowtail pupae, however, the pupae of parnassians are formed in litter on the ground, in a scraggly cocoon. Some authors have referred to this trait as primitive, but it is better understood as a perfectly good adaptation to severe mountain environments. Parnassians also possess a unique feature—the sphragus. When females emerge, the males have already been on the wing for a week or more, and mating occurs as soon as a female has been detected, with no courtship niceties whatsoever. The male tackles his victim, on the ground or in the air, mates with her and deposits a hardened structure on her abdomen that prevents her from mating again, but still allows her to deposit eggs and waste. This structure is called the sphragus.

There are between three and five species of parnassians in North America, but in B.C. there are only three. What to call them is another matter, but still there are clearly three.

Rocky Mountain Parnassian

Parnassius smintheus

Wingspan: 45–70 mm

In most butterfly books, you will find this species referred to as the Phoebus Parnassian (*Parnassius phoebus*). The consensus has long been that the parnassians in North America with striped antennae are all members of a highly variable species that also occurs in Asia, on the other side of the Bering Sea (which was, thousands of years ago, the Bering Land Bridge). The variability of these butterflies led early collectors to give a separate name to each and every different-looking lineup of cabinet specimens, and even now there is little agreement on the names that should be applied to differently marked geographic races. Part of the difficulty in delineating races stems from the fact that within any given subspecies, it is relatively easy to find a few butterflies here and there that look like they belong to some other subspecies. In 1898, lepidopterist W.J. Holland wrote, "Any lively boy, collecting on a ten-acre lot in Montana, when *P. smintheus* is on the wing, can turn up almost all of the so-called subspecies and 'forms,' which now burden our lists, and no doubt many others..." (*The Butterfly Book*, pp. 310-311). Notice that way back then, he was calling this species *P. smintheus*.

In 1994, the prominent Pacific Northwest lepidopterists Jon Shepard and Thomas Manley reclassified the Phoebus Parnassians of North America, based primarily upon examination of the microscopic structure of their eggs. They decided that true Phoebus Parnassians (*P. phoebus*) live only in Asia, Alaska and the Yukon; Rocky Mountain Parnassians (*P. smintheus*) extend from the Yukon down to California; and Sierra Nevada Parnassians (*P. behrii*) live only in the Sierra Nevada mountains of California. This arrangement has been widely accepted by lepidopterists, and I am pleased to follow it here. However, it should be kept in mind that the differences that Shepard and Manley used to delineate the three species are only clues, and not conclusive proof. No doubt, their classification will be carefully tested by other scientists, and if some day the books go back to calling them all Phoebus, don't say I didn't warn you.

For Phoebus Parnassians, open areas high on mountains are like islands surrounded by uninhabitable forests and valleys. In nearby Alberta, ecologist Jens Roland and his students have been studying the movements of these butterflies within and among meadows. They have even perfected a method for attaching tiny radio transmitters to individual butterflies to follow their movements the way mammalogists follow grizzly bears from the air. The results of Jens Roland's studies will have significance for both butterfly enthusiasts and a group of scientists called landscape ecologists, who study the way organisms use complex landscapes.

Also Called: Phoebus Parnassian; *Parnassium phoebus smintheus*.

ID: black-and-white-banded antennae; at least one red spot on most individuals; dark spots on the forewing are black, not greyish; yellow body hair; mated female has a small, grey sphragus.

Similar Species: other parnassians (pp. 36–39) have black antennae and more greyish, dark forewing spots.

Caterpillar Food Plant: stonecrop, especially lance-leaved stonecrop (Crassulaceae: *Sedum lanceolatum*).

Habitat & Flight Season: meadows; open, rocky places and open, sunny forests in the mountains; flies from June to mid-August.

Clodius Parnassian

Parnassius clodius

Wingspan: 53–82 mm

The Clodius Parnassian has long been considered the only true North American member of its genus, but now that we recognize Rocky Mountain and Behr's Parnassians as separate from Phoebus, this is no longer the case. Remember, both Phoebus and Eversmann's Parnassians also occur in Asia. The white colour of parnassians is probably a warning to predators. It has been reported that these creatures take in alkyloid poisons by feeding on such plants as bleeding heart, corydalis and stonecrop as caterpillars, and these chemicals are thought to be retained by the adult butterfly as a defence. The other main group of white butterflies in North America, the whites proper, are also distasteful, and thus white is not simply a "plain" type of colour, it is a highly adaptive one. Where they occur together, Clodius Parnassians resemble day-flying Buck Moths in the genus *Hemileuca*, especially at a distance. Some entomologists have speculated that these moths are also distasteful, and that their colour is a

Similar Species

Eversmann's Parnassian

warning to predators. For a bird, the lesson may be simple: all white and black lepidopterans taste awful.

The caterpillars of the Clodius Parnassian have also been implicated in mimicry because they look a lot like the aptly named Cyanide Millipede (*Harpaphe haydeniana*). The millipede, however, lives only in lower-elevation forests (generally below 1000 m), and is closely associated with Douglas-fir forests. Interestingly, where the Clodius Parnassian lives at low elevations, its caterpillars are black and yellow, like the millipede. At higher elevations, the caterpillars are pinkish grey. The caterpillars of our other two parnassians are also reminiscent of the Cyanide Millipede, but there the resemblance is probably the result of what biologists call "convergence." Black and yellow are common warning colours (they work well to get the message across), and there are only so many ways to produce a worm-like, segmented, black and yellow animal. Hence, both the caterpillars and the millipedes probably evolved their distinctive colour independently.

ID: black antennae, forewing has dark greyish spots and no red spots; wings may be yellowish when fresh. *Female:* large white sphragus; some have hind wing spots forming a band.

Similar Species: *Eversmann's Parnassian* (p. 38): similar, but its range does not overlap with Clodius.

Caterpillar Food Plant: bleeding heart (Fumariaceae: *Dicentra* spp.).

Habitat & Flight Season: meadows, rocky slopes and open woods; flies at lower elevations than the Rocky Mountain Parnassian and earlier in the season, mainly in June; in the northern part of its range, Clodius flies until August, and at lower elevations.

Eversmann's Parnassian

Parnassius eversmanni

Wingspan: 53–61 mm

With its swallowtail-yellow colours, this butterfly makes it easier for the neophyte lepidopterist to believe that the parnassians belong to the swallowtail family. It is difficult to know if the last common ancestor of the Parnassians and swallowtails was yellow or white, but in the presence of a butterfly like the Eversmann's Parnassian, so perfectly in between, this question seems less interesting than I originally thought it might be. The northern haunts of Eversmann's Parnassian have given it an aura of mystery among butterfly enthusiasts, but the story of this species in B.C. is apparently a troubled one. The most accessible location in the province for Eversmann's Parnassians is Pink Mountain, just off

Similar Species

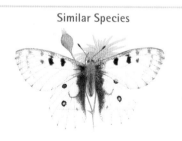

Rocky Mountain Parnassian

the Alaska Highway, and the first specimens from Pink Mountain, collected in 1975, found their way to a European collector who named them as a new subspecies, but on the basis of very sketchy evidence. The "new subspecies" (known as *pinkensis* of all things) was never accepted in scientific circles, but among Eurasian collectors the new name created a demand for specimens. As a result, a number of out-of-province amateurs converged on Pink Mountain in the late 1980s and took every Eversmann's Parnassian they could find. Since then, reports suggest that the population has been but a shadow of its former glory, but it is recovering. Over a century ago, lepidopterist W.J. Holland warned that things like this could happen, with these remarks: "The situation reminds me of nothing more than the celebrated 'tulip-craze,' which once took possession of Holland. This form of nomenclatorial insanity has thus far not deeply infected lepidopterists on this side of the Atlantic, though there are symptoms of the disorder in some quarters" (*The Butterfly Book*, p. 309). The tulip-craze involved frenzied investment, during the 1600s, in bogus new varieties of tulips, which proved to be nothing more than diseased flowers.

ID: black antennae; forewing has dark greyish spots and no red spots. *Male:* yellow ground colour. *Female:* white or pale yellow ground colour; central red or black spot connect to the spots along the trailing edge of the wing, forming a band; mated female has large, white sphragus.

Similar Species: other parnassians are rarely yellowish; *Rocky Mountain Parnassian* (p. 34): lives in the same range.

Caterpillar Food Plant: corydalis (Fumariaceae: *Corydalis pauciflora*).

Habitat & Flight Season: open areas at high elevations; flies in June and July.

Swallowtails Proper
(Subfamily Papiloninae)

Unlike the parnassians, swallowtails are familiar butterflies. They are large, most are brightly coloured in yellows and blacks, and they have the alluring habit of fluttering while they feed, making them delightfully nimble and elusive. All of our species belong to the genus *Papilio*, the members of which generally have hind wing "tails," as well as a downward bend in the hind wing where it meets the body (leading some people to call them the "fluted swallowtails"). Unlike their rough-and-tumble relatives, the parnassians, swallowtails engage in more sophisticated courtship routines, facilitated by both behavioural displays and chemical perfumes.

Swallowtails are also distinctive as larvae, with two-forked, foul-smelling osmeteria like those of parnassian caterpillars. Swallowtails generally begin life looking like bird droppings (and, according to May Berenbaum, at the University of Illinois, tasting like them too; the white colour in both indicates the presence of uric acid). They finish their caterpillar stage looking like short-bodied snakes, complete with fake eyes and the shape of a serpent's head. Not all swallowtail caterpillars look like snakes, however, and those of the Old World Swallowtail group are warningly banded in black, orange, yellow or pink (according to lepidopterist Jim Scott, possibly mimicking the caterpillars of the Monarch butterfly). All swallowtail larvae, however, live under threat from large, black, parasitic wasps in the genus *Trogus*, a genus of ichneumon wasp. Anyone who rears swallowtail caterpillars will eventually meet their share of *Trogus*.

Swallowtail pupae, which are usually either brown or green, are attached to twigs or other substrates by the end of the abdomen, head up, and supported by a simple silk belt around the midriff. The presence of brown and green pupal morphs within the same species may have resulted from the need to remain camouflaged on both brown and green substrates, without the caterpillar having the ability to choose which colour it might become.

Anise Swallowtail on dandelions

Within the swallowtail fauna of British Columbia, we can distinguish two groups of closely related species. The first of these is called the Old World Swallowtail group (or the *Papilio machaon* species group), named for its most familiar member, the Old World Swallowtail. This species was the first butterfly to be officially described by Carl Linnaeus (the founder of biological classification), and thus it is the "type species" of not only its species group, but also its subgenus (in Canada, the subgenus is *Papilio*), genus, family and superfamily. It is, therefore, the central name in all of butterfly classification and the only name that we can expect to remain in existence, no matter how many others are changed.

You might think that this would make the old world swallowtails a well-understood group of butterflies, but historically nothing could be further from the truth. Unless this is the only butterfly book you own, you will undoubtedly find other names in use for the B.C. members of this group, but here I am following the conclusions of Felix Sperling, a good friend for whose tenacity and honesty I have the utmost respect. After thoroughly examining these butterflies in terms of their body structures and wing patterns, Felix looked at enzyme variations and mitochondrial DNA patterns, resulting in a study that I consider one of the truly great triumphs in modern butterfly systematics. In my opinion, if the six species in the Old World Swallowtail species group have not given up their secrets to Felix Sperling, then all other butterfly systematists should pack up their offices and look for new careers. (But don't worry—they did, so the systematists should keep on doing what they love.)

Members of this group of butterflies live marvelous lives. After they emerge, males fly to the tops of nearby hills and ridges, where they congregate in areas with plenty of nectar flowers (imagine, these are the most scenic places in the province!). The males compete for access to these prime hilltops, for this is where females come to meet their mates. Males that are not engaged in driving off other males will wait patiently on the ground for a female to pass by, after which they pursue the female to a nearby grassy

Introduction to the Old World Swallowtail Group

The Old World Swallowtail Group

area and court her. Recently mated females then abandon the hilltops and retreat to lower ground in search of food plants on which to lay their eggs.

Compared to members of the Tiger Swallowtail species group, members of the Old World group are a bit smaller, with more extensive dark wing markings in general. They also lack the thin black line that bisects the hind wing of the Tigers.

Old World Swallowtail—the hind wing eyespot is not always easy to see.

Old World Swallowtail
Papilio machaon
Wingspan: 60–100 mm

This species was named for Machaon, the physician son of Asclepius, the Greek god of health. Calling it the Old World Swallowtail pays homage to the European origins of its name. Over much of Europe, it is the most common swallowtail. In fact, the English simply call it the "Swallowtail," and distinguish it only from the "Scarce Swallowtail" (*Iphiclides podalirius*). From a Canadian perspective, it helps to remember that Europeans can't rightly claim this species as their own because the group as a whole probably originated in North America, and later spread across Beringia to Asia. In fact, owing to its widespread range, all sorts of people like to think this is their swallowtail. To the south, Oregon named it as their state insect a few decades back. Some proud lepidopterists there still insist it is a completely

Similar Species

Anise Swallowtail

separate species *(Papilio oregonius)* rather than "just" a subspecies *(P. machaon oregonius,* as if taxonomic rank has anything to do with the value or magnificence of a butterfly!). And sadly, throughout the Pacific Northwest, dams have flooded much of this butterfly's habitat.

In British Columbia, the Old World Swallowtail is represented by three subspecies; *oregonius* in the south, *aliaska* in the north and *pikei* in the Peace River region. The differences among them are slight, and most people will encounter *oregonius* long before they find the other two. Still, *aliaska* is a striking butterfly, with a hind wing eyespot that is entirely red in the centre, with no black pupil. It is relatively small and stout-winged. *Oregonius* has a distinct dark pupil in the eye spot, and is relatively large and long-winged. *Pikei,* named for Albertan lepidopterist Ted Pike, is a good intermediate between the two, although this is, of course, an oversimplification.

The larvae of the North American subspecies of the Old World Swallowtail are the only swallowtail caterpillars in the world that feed on members of Compositae (the daisy family of plants). Interestingly, wild tarragon contains a chemical called anisic aldehyde, which is also found in many members of the carrot family—a more typical food for this group of swallowtails. The larval stage lasts about a month, and the pupa lasts only about two weeks in the summer generation and many months (overwintering) in the spring generation. If conditions are too dry when the time comes for emergence, some pupae will wait another year or two and try again.

Also Called: Oregon Swallowtail or Baird's Swallowtail; *P. oregonius* or *P. bairdii.*

ID: check for the oblong hind wing pupil in the eyespot that is connected to the black margin around the spot; hairs at the base of the legs are yellow; where there are two generations, spring butterflies are a bit smaller and darker than their summer descendants; sexes are similar.

Similar Species: *Anise Swallowtail* (p. 46): has eyespot pupil surrounded by orange.

Caterpillar Food Plant: wild tarragon, also called dragonwort (Compositae: *Artemisia dracunculus*).

Habitat & Flight Season: rolling or mountainous country, often on sunny, open hilltops with nectar flowers. In the south, there are two generations, one emerges in April and May, the other in July and August; northern populations have only one generation per year and fly mainly in June and July.

Anise Swallowtail
Papilio zelicaon
Wingspan: 60–95 mm

♂
♀

Many butterfly experts have called this the most common swallowtail west of the Rockies, and in British Columbia they are probably correct. The Anise Swallowtail is generally more abundant than the Old World, and it probably outnumbers the members of the Tiger Swallowtail group as well. The name *zelicaon* probably comes from the same root as *zealous*. The name was likely coined in such a fashion that it rhymes with *machaon*, for ease of remembrance (as if!). In any event, they are certainly zealous butterflies. One study showed that some males can live as long as a month. When

Similar Species

Old World Swallowtail

they were taken from their original hilltop, males could sometimes relocate it from a distance of five kilometres or more.

To the east and south of British Columbia, a rare, mostly black, colour morph of the Anise Swallowtail is sometimes found. This was once considered a separate species (as was just about every other minor variant in the group), called the Nitra Swallowtail. Nitras were highly prized by collectors in the days when I was first learning my butterflies. In British Columbia, however, the nitra morph is unknown, and for good reason. It turned out to be the result of a hybridization between Anise Swallowtails and Black Swallowtails (the common eastern member of the Old World group). The genes for blackish wings became established in those populations of the Anise that had been in contact with the Black Swallowtail, and when these rare genes expressed themselves—voilà—Nitra!

Again, the English name of the species deserves comment. Sweet fennel is a plant that smells a bit like licorice, and it is a food plant for this species. In the old days in California, some people called sweet fennel "anise," so it is from this historical accident (which the botanists have corrected in their own books, by the way) that the butterfly name arose. Some people feel that this is good reason to change the name back to "Zelicaon Swallowtail," or "Western Swallowtail," but to me the story of the name is part of what makes this species unique. The attractive green, black and yellow caterpillars of the Anise Swallowtail are relatively easy to find on the foodplants, and a pleasure to rear in captivity. When full grown, they become lovely brown or green pupae, attached to sticks by a thin belt of silk around their midline. These will need to be cooled over the winter, or they will not hatch the following summer.

Also Called: Zelicaon Swallowtail.

ID: black pupil in the hind wing eyespot is round and not connected to the black border; black hairs at the base of the legs; sexes are similar.

Similar Species: *Old World Swallowtail* (p. 44).

Caterpillar Food Plants: various members of the carrot family (Umbelliferae), including angelicas (*Angelica* spp.), cow parsnip (*Heraclium* spp.) and desert parsley (*Lomatium* spp.).

Habitat & Flight Season: sunny habitats, including gardens and parks, valleys, slopes and hilltops; flies from June to August.

Indra Swallowtail
Papilio indra
Wingspan: 65–85 mm

♂ ♀

This is the smallest and darkest of our swallowtails, and also the most divergent member of the Old World group. In other words, it is much less closely related to the other two than they are to each other. In British Columbia, the Indra Swallowtail (not "Indira" as many people mistakenly say—it has nothing to do with the former Prime Minister of India) is represented by only one race (*P. indra indra*). Farther south, there are another ten or so races currently recognized, and plenty of others that have been named and then "sunk" (i.e., synonymized). It is one of the most variable of all swallowtails, a fact that is sometimes lost on us northerners.

Like other members of the Old World group, Indra males are hill-toppers, and they have a reputation for holding especially pugnacious contests for the best spots on the hill. For this reason, as well as others (the life of a butterfly is hard!), most males become worn and tattered soon after they emerge from the pupa. Speaking of which, pupae of this species, like others in the Old World group in B.C., are capable of waiting out a second season if conditions are not to their liking.

The larvae of the Indra Swallowtail are also unusual, banded with pink and black, and sometimes with a few orange spots as well.

ID: an easy butterfly to recognize in B.C.; small, with short tails and a very straight yellow band through the forewings; the all-black abdomen has yellow splotches on its sides near the tip; sexes are similar.

Similar Species: none.

Caterpillar Food Plant: carrot family plants, especially desert parsley (Umbelliferae: *Lomatium* spp.).

Habitat & Flight Season: found only in Manning Provincial Park, near Allison Pass; flies in late June.

Tiger Swallowtail Group

Western Tiger Swallowtail with missing tails, possibly snipped off by a bird.

The remaining species of British Columbian swallowtails fall into the Tiger Swallowtail group, which specialists have called the *Papilio glaucus* species group, after the Eastern Tiger Swallowtail, its first-described member. These are large, mostly yellow (or white) butterflies with great flopping wings and a commanding presence at nectar flowers. Typically, they also have a thin, black line running through the middle of the hind wing, extending the innermost black stripe on the forewing, a feature not seen in the Old World Swallowtail group.

In some books, you will see the members of this group placed in the genus *Pterourus*. Here, as in most books published in the last few years, *Pterourus* is considered a subgenus of *Papilio*, in which the Tiger Swallowtail group is nested. You may also see the genus name *Euphoeades* used—evidence of yet another attempt to split the large genus *Papilio* into smaller units (although *Euphoeades* was apparently not a useable name, according to the official rules). In contrast to this historical trend toward splitting off the Tiger group from the rest of *Papilio*, in older books you will also find a classic example of taxonomic lumping. Both the Canadian and Western Tiger Swallowtails were once often considered subspecies of the Eastern Tiger Swallowtail (which was, of course, considered *the* Tiger Swallowtail). Like the Old World group, these butterflies did not give up their secrets easily, and like them, the best research has come in the very recent past, with new and better techniques for determining who is related to whom, and in what capacity.

Western Tiger Swallowtail

Papilio rutulus

Wingspan: 75–105 mm

I n B.C., this is a southern species. To the north, it is replaced by the Canadian Tiger Swallowtail, a close relative. In fact, hybrids between the two species can be found in south-central B.C., showing a mix of the parental characteristics. Most butterfly experts consider the Western Tiger Swallowtail to be the western equivalent of the Eastern Tiger Swallowtail, and indeed they seem to replace one another along a diagonal line through the middle of the United States. To the experienced eye, this looks like evidence that the two species were once one, and that they recently diverged from their common ancestry; however, molecular studies have shown that the Western Tiger Swallowtail is more closely related to the Pale Swallowtail than to the Eastern Tiger, suggesting that the traditional view might be mistaken. It will be interesting to watch this area of research, and see how the evidence from wing pattern, geographic

Similar Species

Canadian Tiger Swallowtail

ranges and molecular systematics is interpreted in light of one another. Oh, and to make things more complicated, the Western Tiger Swallowtail can also hybridize with the Two-tailed Swallowtail and the Canadian.

Encountering a Western Tiger Swallowtail while wielding a net is like meeting a professional athlete at a cocktail party, and foolishly deciding to test your mettle. One moment, you are confidently addressing

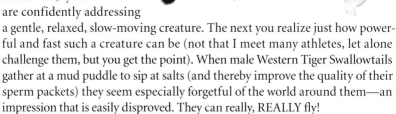

a gentle, relaxed, slow-moving creature. The next you realize just how powerful and fast such a creature can be (not that I meet many athletes, let alone challenge them, but you get the point). When male Western Tiger Swallowtails gather at a mud puddle to sip at salts (and thereby improve the quality of their sperm packets) they seem especially forgetful of the world around them—an impression that is easily disproved. They can really, REALLY fly!

Whereas the Anise Swallowtail is probably our most common species of swallowtail, the Western Tiger Swallowtail commands more attention because it is much bigger and more obvious. The caterpillars, on the other hand, are not obvious at all; they spend much of their time in a rolled up leaf, saving their snake-head colour for those special moments that really warrant such things. The variety of plants that Western Tiger Swallowtail caterpillars eat make them some of the least fussy among the swallowtail larvae.

ID: a large, primarily yellow swallowtail, with only one prominent tail; orange markings are seen only on the side of the tail nearest the body; sexes are similar.

Similar Species: *Canadian Tiger Swallowtail* (p. 52): orange markings on both sides of the tail base on the underwing.

Caterpillar Food Plants: willows, poplars, birch and cherry (Salicaceae: *Salix* spp., *Populus* spp.; Betulaceae: *Betula* spp.; Rosaceae: *Prunus* spp.).

Habitat & Flight Season: valleys along river and stream channels, as well as in parks and gardens; flies in May and June at lower elevations, June and July higher up.

Canadian Tiger Swallowtail

Papilio canadensis

Wingspan: 70–105 mm

This is the swallowtail I know best, so I may well be biased toward it. It's not the biggest, the brightest or the rarest, but hey, it's Canadian, right? Well, sort of. For decades this swallowtail was considered a race of the Eastern Tiger Swallowtail that mysteriously did not have a dark morph among the females. (In the true Eastern Tiger Swallowtail, some females mimic the dark, distasteful Pipevine Swallowtail, *Battus philenor*.) Reading the older books, it is amazing how our delightful Canadian Tigers were characterized. Consider the words of W.J. Holland, in 1898: "Such dwarfs [what he called "variety *canadensis*"] are also common in the early spring in Pennsylvania and

Similar Species

Western Tiger Swallowtail

West Virginia. They reflect the effects of wintry cold upon the chrysalis" (*The Butterfly Book*, pp. 318-319). Baloney!—as if being a bit smaller demanded an explanation, bigger naturally being better.

The Canadian Swallowtail sometimes hybridizes with the Western Tiger Swallowtail, discussed under the latter species' treatment. I should also mention that some older female Canadian Tiger Swallowtails seem to have white and black front wings and yellow and black hind wings. These, however, are simply faded from basking in the sun, and are not genetically different from the others.

The caterpillars of this species, like the last, are renowned for feeding on a variety of plant groups, and are interesting in that the pupa usually overwinters in leaf litter rather than in open air. Under an insulating blanket of snow, temperatures can remain close to the freezing point, even when things get really, really cold above the snow.

Also Called: *P. glaucus canadensis*.

ID: generally the smallest member of the Tiger Swallowtail group; orange spots on hind wing are between the body and tail and along the wing margin above the tail. *Male:* smaller than female; darker yellow ground colour as opposed to orange-yellow or pale yellow; otherwise, sexes are similar.

Similar Species: *Western Tiger Swallowtail* (p. 50).

Caterpillar Food Plants: willow, poplar, cherry, ash and crab apple (Salicaceae: *Salix* spp., *Populus* spp.; Rosaceae: *Prunus* spp., *Malus sylvestris*; Oleaceae: *Fraxinus* spp.).

Habitat & Flight Season: gardens, open forests, along forest edges, along stream channels and woodland roads; flies from May through July.

Pale Swallowtail

Papilio eurymedon

Wingspan: 75–110 mm

Male Pale Swallowtails, like their close relatives, are regularly found on wet mud, sipping at the salts that they require for the production of high-quality sperm. This species is also well known for its attraction to thistles as nectar flowers, and for its habit of cruising along high elevation ridges and roadways, in perpetual search of potential mates.

To me, this butterfly always looked awkward in the field guides, as if something were wrong with it. Now that I've had a few decades to think about it, I suspect that what influenced me was the name. Somehow, "Pale Swallowtail" suggested "pale imitation" or "pale by comparison," and I saw this species as an inferior version of its brighter relatives. I suppose that if the others had been named the "yellow-stained swallowtails" I would have grown up with the opposite bias. In truth, of course, they are all marvelous in their own ways, and the look of a Pale Swallowtail on the wing is something quite exotic and unlike any of the other swallowtails of British Columbia.

If anything, the broad dark markings of this butterfly make it the darkest of all the tiger swallowtail group members in British Columbia. When you see one of these butterflies on the wing, the combination of generally dark colour, lack of yellow, and habitat (especially hilltops) all make it fairly easy to identify.

The name *eurymedon*, speaking of names, means "broad guardian," and was originally the name of a Greek river, as well as a king of giants. Thus, it continues the tradition of using historical or mythological Greek scientific names for this group, in contrast to the more descriptive names *multicaudata* (many tails) and *glaucus* (grey). If scientific names were made to be descriptive (they are not—they are merely labels for named groups of organisms), then perhaps the Pale Swallowtail should have been the one to be called *glaucus*. The larvae of this species are typical of the group, although they generally live communally rather than alone. As well, their pupae are almost always brown and not green.

The Pale Swallowtail is not very common anywhere, even where it is an expected member of the fauna. On the other side of the Rocky Mountains, in extreme southwestern Alberta and the Waterton Lakes/Glacier National Park area, this is a very rare butterfly indeed. For someone like me, who learned butterflies in Alberta, it always strikes me that British Columbians are super fortunate to have this species so widespread and reliable in the province.

Of all the members of the tiger swallowtail group, this is the species that is least inclined to hybridize with its close relatives. If, however, you see a very pale female Canadian Tiger Swallowtail, don't immediately assume that I am wrong here—sunlight sometimes fades the yellow, especially on the forewings of Canadian Tigers, to the point where beginners mistake them for Pale Swallowtails.

ID: a large, pale swallowtail with a creamy white ground colour; broad black wing borders and stripes; sexes are similar.

Similar Species: none.

Caterpillar Food Plant: trees and shrubs in the families Rosaceae and Rhamnaceae, notably cherry (*Prunus* spp.), crab apple (*Malus sylvestris*) and tea bush (*Ceanothus* spp.).

Habitat & Flight Season: valleys, stream courses, slopes and hilltops, mountainous country; flies from late April through July, depending on location and weather.

Two-tailed Swallowtail
Papilio multicaudata
Wingspan: 85–120 mm

In British Columbia, the Two-tailed Swallowtail is one of three giants in its family. In the southern United States, however, it is a truly huge butterfly, even bigger than the Monarch (*Danaus plexippus*), and the so-called Giant Swallowtail (*Papilio cresphontes*). Ernst Dornfeld, author of *The Butterflies of Oregon* (1980), called the Two-tailed Swallowtail "lordly," and in so doing he most likely echoed the words of the grand old man W. J. Holland, in 1896. Holland referred to the Eastern Tiger Swallowtail as the "Lordly Turnus," when its scientific name was *Papilio turnus*, not *P. glaucus*. The fact remains that all members of the Tiger Swallowtail group are indeed lordly, and we are

Similar Species

Western Tiger Swallowtail

lucky to have four of them in British Columbia.

Now, let me dwell for a moment on a more important issue. How many tails does this butterfly have, really? Two? Well, have a close look. One of the tails is surely longer than the other, and both are simply exaggerations of the scalloped wing margin that extends along the outside edge of the wing. Also notice that the innermost such bump is almost as long as the second "tail" even if it is broader at the base. With this in mind, have a look back at the Western Tiger Swallowtail and the Canadian Tiger Swallowtail. They, too, have a second, albeit miniscule, tail. This explains, in part, why the scientific name of the Two-tailed Swallowtail means "many tailed." Why not call it the Three-tailed Swallowtail? Well, because there is a Mexican species (*Papilio pilumnus*) with three (more or less) tails, that occasionally reaches the southern U.S.

You'll notice the scientific name of this species is spelled *multicaudata*. Although I use this spelling, rather than *multicaudatus*, I consider the former incorrect, and the reason is simple, and, I think, interesting. All scientific names must conform to the rules of Latin grammar, and in Latin some nouns are masculine, some are feminine. If the genus of a butterfly is masculine, so must be the specific epithets of its members. *Papilio* is masculine, and so is *multicaudatus*, while *multicaudata* is feminine. Notice that the –*us* ending is also present on *glaucus* and *rutulus*. However, lepidopterists, as a group, have informally decided to ignore the official rules, and use the original spellings of names rather than those that are technically correct. I find it kind of fun to work my way through species lists and try to spot the names that do not correspond properly in gender. And I also enjoy looking through butterfly guides for far-away places, and comparing their most impressive species with ours. If you haven't already done this, I think you will find that few butterflies anywhere in the world are more dazzling than the Two-tailed Swallowtail. At some level, it simply becomes a matter of personal preference and "what's your favourite colour."

Also Called: *P. multicaudatus.*

ID: large and mostly yellow with two tails (one much longer than the other), rather than one; black stripes on the forewing are relatively narrow, compared to other members of the Tiger Swallowtail group; sexes are similar.

Similar Species: *Western Tiger Swallowtail* (p. 50): has only one tail.

Caterpillar Food Plants: willow, poplar, cherry, saskatoon, ash and crab apple (Salicaceae: *Salix* spp., *Populus* spp.; Rosaceae: *Prunus* spp., *Malus sylvestris;* Amelanchier alnifolia; Oleaceae: *Fraxinus* spp.).

Habitat & Flight Season: valleys near water, as well as gardens and parks; flies in May and June and sometimes with a partial second brood in September

Whites and Sulphurs
(Family Pieridae)

Personally, I would be happier if we could call this family the whites and yellows because most members of the group are exactly that. The Pieridae in British Columbia falls into two easily recognized subfamilies. The adults are generally medium-sized, with rounded wings and neatly marked patterns. The whites, marbles and orangetips are generally white (some have orange wing tips or greenish marbling on the underside), and the sulphurs are generally yellow (and in some places they are indeed called "yellows"). Their larvae are for the most part quintessential caterpillars, without spines or other outgrowths, and generally greenish or yellowish in colour. Pierids are usually patrolling butterflies, seen flitting about over fields and open areas. They fly well, but often in a somewhat erratic fashion.

Male Checkered White (left)
Christina Sulphur, showing just a hint of orange on the forewing (above)

Whites, Marbles and Orangetips
(Subfamily Pierinae)

Olympia Marble, a species that has not yet been found in British Columbia.

Whites are white for a good reason: their colours tell bird predators that these butterflies taste bad. For this reason, they are also bold, and flutter about without much apparent fear for their safety. This may explain why the whites are, in general, slower on the wing than the sulphurs, their close relatives in the white and sulphur family. White, yellow, black and green are the typical colours of whites, and they rarely exhibit any of the orange or red colours that are common in other families. White scales owe their origins to waste chemicals that are saved within the body of the caterpillars and then deposited in the wings. Black is formed from melanin, just as it is in humans, and yellow pigments are also of the typical "carotenoid" sort found in a variety of animals (and, not unexpectedly, in carrots).

Stella Orangetip

Green, however, is formed from interspersed black and yellow scales that, oddly enough, give the impression of being a greenish colour when viewed from a distance.

Whites generally hibernate as pupae, and are typically butterflies of spring, or at least the beginning of the butterfly season, which comes later at higher altitudes. That is, except the Pine White, which overwinters in its egg stage and appears in late July. A few whites are multi-brooded and appear on the wing two to three times or more throughout the butterfly season, one brood overlapping with the next.

It may help to think of the whites in three subgroups, corresponding to the three genera in our fauna. The first has only one member, the Pine White. The second is made up of the *Pontia* or "checkered" whites, which have a characteristic black-speckled wing pattern. The third includes the mustard white group and the Cabbage White, and is the most familiar by far.

Like the whites, the marbles and orangetips feed on mustard family plants as caterpillars (generally eating the flower parts and not the leaves), and are protected as adults by the toxic mustard oils that are deposited in their white wings. *Arabis* (rock cress and its relatives), is a very important food plant genus. The marble genus *Euchloe* means "true spring green," referring to the greenish marbling on the underside of the hind wings. Like whites, however, their green colours are not true, but are formed from black and yellow scales. *Anthocharis,* the genus name for orangetips, means "flower grace," and is more accurate as well as more poetic. To most butterfly people, the marbles and orangetips are seen as a cut above the whites on the wish-I-could-see-one scale, with a bit more snap to their wing patterns and a certain element of comparative rarity, at least in most places.

Pine White

Neophasia menapia

Wingspan: 45–55 mm

♀

The Pine White is an unusual butterfly. For one, it spends most of its time in the treetops, especially in the afternoon. Males visit flowers at ground level in the morning, but the females seem rarely to do so. For the rest of the day, Pine Whites flutter and parachute high among the Douglas-firs, seeking out mates, egg-laying sites and perhaps additional sources of nutrition. The caterpillars develop on needles in the treetops, and then descend on silk threads to pupate near the ground. In the early days of North American butterfly study, John A. Comstock (the author of the classic, *Butterflies of California* [1927]) discovered that adult male Pine Whites would fly down from the forest canopy to investigate, albeit briefly, a pinned specimen of their own species, or even a white piece of paper.

There was an outbreak of Pine Whites on Vancouver Island in 1961, but since that isolated incident, it would be inaccurate to call this species a forest pest. Most sources say that the damage caused by occasional outbreaks of Pine Whites is usually part of the aftermath of bark beetle infestations, which weaken the trees and increase their susceptibility to damage by caterpillars.

Still, there have been instances in Idaho where Pine White larvae were responsible for the deaths of approximately one-quarter of all trees within an outbreak area. Most butterfly books will tell you that this species is primitive, but few

explain what this term really means. It means that the Pine White and its close relatives are, in a general sort of way, more like the ancestor of all whites than are any of our other species. They share this distinction with the Chiracahua White (*Neophasia terlootii*), a denizen of the mountains of southeast Arizona and adjacent Mexico, and the Dart Whites (*Catasticta* spp.) of Mexico, Central and South America. In turn, with a good imagination you should also be able to see similarities between Pine Whites and Parnassians, and even greater similarities between Dart Whites and more "primitive" members of the swallowtail family (for example, the genus *Baronia*). Indeed, in most classifications, the swallowtails, and the whites and sulphurs, are seen as two very closely related groups, distinct from the rest of the butterflies.

ID: distinctive up close; white with black leading edge along the base of the forewing, as well as dark outer border with light spots inside; slower and higher flight style than other whites, or white-coloured sulphurs. *Male:* no dark outer border on hind wing; thin veins on underside of hind wing. *Female:* dark outer border on both pairs of wings; thick veins on underside of hind wing; sometimes orange markings on hind wing underside.

Similar Species: other whites and white-coloured sulphurs.

Caterpillar Food Plants: Douglas-fir and some pines (Pinaceae: *Pseudotsuga menzeizii* and *Pinus* spp.).

Habitat & Flight Season: coniferous forests, especially with Douglas-fir; flies in late July in the lowlands, and persists through September in cooler regions.

Western White
Pontia occidentalis
Wingspan: 35–50 mm

For a long time, in the minds of butterfly experts, there was no such thing as a Western White. This butterfly was originally thought to be a subspecies of the Checkered White, but in the early 1960s, V.C.S. Chang and William Hovanitz demonstrated that the two should be considered separate species. Most naturalists, however, had to wait for the monumental butterfly guide by William Howe, published in 1975, to discover that these two are really "true sibling species." Sibling species do not interbreed, are very difficult or impossible to tell apart and live in the same places, at least over part of their respective geographic ranges. They are often evolutionary siblings as well; it is typical for such pairs of species to have evolved from a common ancestor, with the

Similar Species

Becker's White

Checkered White

Spring White

ancestral population, at some point, being split into two. During the period of separation, the two new lineages diverged genetically, such that when they were reunited they could no longer interbreed. In the case of the Western White, however, the story is not entirely so simple. It has been shown that Western Whites can interbreed, in a weak fashion, with Peak Whites (*Pontia callidice*), a species of central and Mediterranean Europe. For this reason, some feel that our Western Whites should actually be considered a subspecies of the Peak White, not the Checkered White. This idea is not likely to gather much support, at least in the short run, because butterfly taxonomists are now much more interested in recognizing differences among butterflies than in recognizing similarities. As well, enzyme studies have shown that Western and Peak Whites are quite distinct, and not likely to warrant a single species name.

ID: the most familiar of the four dark-spotted *Pontia* whites; thin, yellow veins on the hind wings are outlined in a less contrasting grey or green; adults appear in the spring; those at higher elevations are smaller and darker than average; dark markings are more extensive and more numerous on the females.

Similar Species: very similar and difficult to separate from *Checkered White* (p. 66); *Becker's White* (p. 68) and *Spring White* (p. 70) have bright yellow veins on the hind wing underside, outlined in contrasting black or green. *Becker's White* is sparingly and crisply marked, with a large, dark forewing cell spot. The *Spring White* is small and dark with no white inside its narrow forewing cell spot.

Caterpillar Food Plants: mustard family plants (Brassicaceae), including *Sisymbrium* spp. and stinkweed (*Thlaspi arvense*).

Habitat & Flight Season: open habitats, preferring natural to agricultural or weedy settings, extending to the alpine; one, two or three broods per season; in the north and the mountains, it is a summer butterfly, but flies throughout the butterfly season in warmer lowlands; does best in warm, dry years.

Checkered White

Pontia protodice

Wingspan: 35–50 mm

♀

In 1904, the American lepidopterist John H. Comstock (of Cornell University, not to be confused with his lepidopterist contemporary John A. Comstock—those were confusing times) and his wife Anna wrote in *How to Know the Butterflies* that "the Checkered White is distributed over the whole United States, though its natural home is in the Mississippi Valley" (p. 75). This sentiment persists to this day, as if this species' appearance elsewhere is in some sense unnatural. Also called the Common White (but not very commonly), this is one of many suicidal butterflies that invade the province each year. To someone not used to the ways of insects, it seems unfair, or at least unwise, for the butterflies to fly north each summer only to doom their offspring to death by frost in the fall.

Similar Species

Becker's White

Western White

Spring White

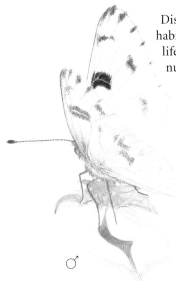

Dispersal into a variety of suitable and unsuitable habitats, however, is part and parcel of the "weedy" lifestyle of such butterflies, where females lay vast numbers of eggs, most of which have no chance for survival whatsoever. Once you realize this, you have to admit that all living things are in some sense suicidal, by producing more young than will survive, and spreading their progeny hither and thither. That *some* survive is all that is needed to continue this blind tradition. And if global warming does indeed change the face of southern B.C., someday the Checkered Whites will survive here year-round.

John H. and Anna Comstock also wrote that the Checkered White "was once very abundant, but the introduction of the European species [the Cabbage White] has imposed upon the Checkered White a checkered career, and it is now rarely taken" (*How to Know the Butterflies*, p. 75). I discuss the effect of the Cabbage White more fully under that species. It may be that the reason for this pattern has only come to light recently, and it may well have to do with the introduction not of the Cabbage White, but of a parasitic wasp that was intended to control the Cabbage White. The wasp, a braconid named *Cotesia glomerata*, may indeed parasitize Checkered White caterpillars when it gets the chance.

ID: very similar to the more common *Western White*; male has more black markings on the wings than the female.

Similar Species: *Becker's White* (p. 68) and *Spring White* (p. 70): both are similar. The male *Western White* (p. 64) has fewer dark markings than the Checkered White; the female Checkered White has slightly browner wing markings then the Western, but this is not always apparent.

Caterpillar Food Plants: mustard family plants (Brassicaceae).

Habitat & Flight Season: weedy and disturbed places, lowland meadows; they arrive in the Kootenay region in mid-summer from the U.S.; they lay eggs and produce young, but most pupae do not survive winter in B.C.

Becker's White
Pontia beckerii
Wingspan: 35–50 mm

♀

Also known as the Great Basin White and the Sagebrush White, this butterfly is sometimes considered a subspecies of the Small Bath White (*Pontia chloridice*), which lives in Mediterranean Europe. In general, there is a strong trend away from using European species names (which almost always predate our own) for North American animals, even when the two seem clearly connected by recent common ancestry. At the very least, it avoids the issue of whether we are obliged to follow the European English names as well, but this trend is also symptomatic of the slow but sure separation of

Similar Species

Checkered White

Western White

Spring White

the North American lepidopterist tradition, echoing the evolution of distinctive North American forms.

Among our four species of checkered whites, this one is perhaps the most distinctive in appearance, and probably even more so to the eyes of other butterflies, with strongly reflective wing undersides in the ultraviolet. Becker's Whites are strong fliers, and edgy as well. When you encounter one, you will no doubt have trouble approaching closely, and once you flush one from hiding you might as well go looking for another, at least in most instances. In contrast to the flighty adults, the pupae sit patiently through the winter, looking like a bit of bird poop slightly peeling away from a twig.

ID: large black cell spot with a white centre; bright yellow veins on the underside of the hind wing are thickly lined with green scales; light crescent on underside of the hind wing. The spring and fall generations produce smaller adults than the summer generation. *Male:* white upper surface on the hind wing. *Female:* dark border on the upper hind wing.

Similar Species: *Spring* (p. 70), *Checkered* (p. 66) and *Western Whites* (p. 64): similar, but none share the combination of a large, dark, white-centered cell spot in the forewing and bright yellow veins outlined in green on the hind wing underside.

Caterpillar Food Plants: rock cress (Brassicaceae: *Arabis* spp.) and other mustard family plants; feeds on buds and flowers more than leaves.

Habitat & Flight Season: dry, shrubby grasslands in the southern Interior; two or three broods throughout the warmer months.

Spring White

Pontia sisymbrii

Wingspan: 35–45 mm

With only one brood, which predictably emerges early in the season, the name Spring White is surely appropriate for this species. It is also known as the California White in some books, but I've always felt that naming animals for part, but not all, of their geographic ranges is a mistake. The scientific species name refers to the mustard genus *Sisymbrium*, one of this species' caterpillar food plants. Like so many other insects, the Spring White reminds us of the amazing diversity of life. On the surface, it looks so much like the other three local members of the Checkered White clan, but its biology is distinctly different.

Similar Species

Checkered White

Western White

Becker's White

Why, for example, does the Spring White have only one generation per year, while its relatives go through two or three? Why is the Spring White not migratory, while the Checkered White is strongly migratory? Why is the Spring White's pupa not bird-poopish? The answer lies in recognizing a common fallacy among simplistic versions of evolutionary thinking—that an adaptation in one species is both available and logical for that species' close relatives. It may well be that the genetic potential for being multi-brooded, migration and pupal camouflage simply never arose in the history of the Spring White. Or perhaps for this species, the opposite adaptations conferred an even greater degree of ecological success.

ID: a relatively small, checkered white with a narrow, all-black cell spot on the forewing; thinly black-etched veins on the upper surface; yellow veins with black trim on the under surface. *Male:* lacks the additional spot on the upper forewing, to the rear of the cell spot. *Female:* additional spot on the upper forewing; in southern B.C., female is usually pale yellow, not white.

Similar Species: the Spring White is most easily distinguished from the *Western White* (p. 64), *Checkered White* (p. 66) and *Becker's White* (p. 68) by its narrow, all-dark, forewing cell spot. Spring Whites are generally smaller than other *Pontia*.

Caterpillar Food Plants: rock cress (Brassicaceae: *Arabis* spp.) and other mustard family plants, including *Sisymbrium* spp.

Habitat & Flight Season: dry, open areas in southern Interior; single brood emerges in April to June at low elevations, and usually July at high elevations.

Veined White
Pieris oleracea

Wingspan: about 40–50 mm

This is the Mustard White of eastern Canada and the eastern U.S. and as such, it only enters British Columbia in the north, where it has dispersed over the Rockies and into the northern Interior. It is a clean-looking, neatly marked butterfly and, in its own way, very beautiful. Before the arrival of the Cabbage White and its introduced parasites, the Veined White was a crop pest in its own right, but since the end of the 19th century, this butterfly has retreated to the woodlands and left the crops and gardens to the Cabbage White, its introduced relative. Some butterfly scientists believe that the range of the Veined White originally expanded with European colonization and a variety of edible crops, and that even though this butterfly is now less common than it once was, it is more widespread. As indicated above, in some books it is still called the Mustard White or the Veined Mustard White.

summer generation
♂♀

Similar Species

Arctic White

Also Called: Mustard White; (*P. napi*).

ID: entirely white on the upper wing surfaces; hind underwing is white with dark veins in spring generation, or in cold places; summer generation is yellowish white with no dark veins. Sometimes female has dark spots on the upper forewings, but the female is usually not distinguishable from the male on the basis of wing pattern.

Similar Species: *Margined White* (p. 73) and *Arctic White* (p. 74): have more dark markings on the wings; the Arctic is also smaller.

Caterpillar Food Plants: mustard family plants, including the cabbage (Brassicaceae: *Brassica oleraceae*).

Habitat & Flight Season: flies low to the ground in open areas and in or adjacent to woodlands; generally two generations per year: the spring brood emerges from over-wintered pupae and the summer generation comes along in July.

Margined White

Pieris marginalis

Wingspan: about 40–50 mm

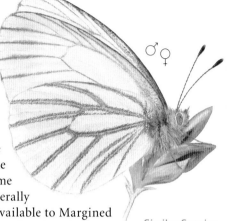

The Margined White is the widespread member of the Mustard White group in western Canada. This species has two broods in southern British Columbia, but only a single brood in the north, where the season is too cool to allow sufficient growth through the season.

Since the arrival of the Cabbage White, opinions on the history of the Margined White have varied. Some feel that the loss of forests has generally reduced the amount of habitat available to Margined Whites, but others see things differently. It has also been suggested that the caterpillars of this species feed best on toothwart (*Dentaria diphylla*), which has become less common, and less well on garden yellowrocket (*Barbarea vulgaris*), a widespread, introduced weed.

Some books still call this butterfly the Mustard White (*Pieris napi*), and some call it the Margined Mustard White (*Pieris napi marginalis*). You may also find the Mustard White referred to as the Sharp-veined White, in contrast to the Diffuse-veined White (*P. virginiensis*), which is more commonly called the West Virginia White. Because diffuse vein scaling distinguishes the Margined from the Veined Whites, this name is not one to miss. And if you think the confusion stops there, some experts divide the British Columbia populations of the Margined White into four separate subspecies, whereas others feel that no more than a single subspecies is represented here.

Similar Species

Arctic White

Also Called: Mustard White; (*P. napi*).

ID: named for a thin, dark margin on the upperwing surfaces. Underwing veins are relatively light, greyish and smudged. In cooler and wetter areas, their dark markings are more extensive.

Similar Species: *Veined White* (p. 72): darker, more crisply marked underwing veins. *Arctic White* (p. 74): smaller and darker.

Caterpillar Food Plants: mustard family plants (Brassicaceae).

Habitat & Flight Season: open areas in forests, from April through August in warmer areas, and mostly July where it flies in only one brood.

Arctic White
Pieris angelika
Wingspan: 35–45 mm

Similar Species

Margined White

Veined White

Here's the good news—this species has been recognized so recently that it has only one English name (and I'm certainly not going to be the first person to mess this up!). Its scientific name honours Ulf Eitschberger's wife, and may still change as the Mustard White group is further studied, but for now, things seem more stable than they have been for decades.

In British Columbia, this is one of those mysterious species of the far north that few people ever see. Its range is confined to the extreme northwest corner of the province. It is more or less separate from the other two Mustard Whites, but the three occur together in a relatively small area around Atlin. There, it is believed, they remain separate species from one another. And luckily, outside this small area, you needn't confront the difficulty of distinguishing the species from one another, side by side.

ID: a smaller Mustard White, with slightly more pointed wings. *Male:* white with thin black veins on the upper surface, wing borders narrow toward wing margin. *Female:* dark dorsal wing veins and borders, borders widen toward margin, other dark markings are prominent.

Similar Species: *Margined White* (p. 73) and *Veined White* (p. 72): generally larger, with lighter wing margins.

Caterpillar Food Plants: not known here, but presumably mustard family plants (Brassicaceae).

Habitat & Flight Season: flies in forest openings, like its relatives, but also on wet tundra; single brood emerges in the first half of summer.

Cabbage White

Pieris rapae

Wingspan: 35–50 mm

♂

The most familiar butterfly in North America, this species is also called the Cabbage Butterfly, the Cabbage Moth and the Imported Cabbage Worm. In Europe, where it comes from, it is the Small White. Cabbage Butterflies first appeared in North America in 1859 in Quebec, and have since spread across the continent, probably reaching British Columbia by the late 1800s. In 1904, lepidopterists John H. and Anna Comstock were already fuming at this species in the eastern U.S., writing "this is not a pretty butterfly… its black-tipped and spotted white wings lend no color to the landscape" (*How to Know the Butterflies*, p. 80).

Able to lay 700 eggs per female, these butterflies quickly displaced their native relatives, although even 100 years ago it seemed apparent that this was not a matter of head-to-head competition. Instead,

Similar Species

Margined White

Veined White

it was widely suspected that the Cabbage White did best in the rapidly expanding crop and garden habitats, whereas the native whites did best in the rapidly shrinking native meadows.

Recently, however, a new idea has come to the fore. It turns out that in 1883, entomologists intentionally released a tiny braconid wasp, *Cotesia glomerata*, in the hopes that it would control Cabbage White caterpillars the way it did in Europe. This is, in fact, one of the earliest examples of what we now call biocontrol. It may be that it suffers from the same problem that plagues most of biocontrol—the accidental decimation of nontarget species. A second wasp, *Cotesia rubecula*, was introduced to New England in the 1960s, and only now are the effects of these two wasps being studied. Roy Van Driesche at the University of Massachussets has obtained evidence that the wasps do not invade forests or forest meadows, and thus the "retreat" of the native whites into these areas may have more to do with parasites than with any preference for this sort of habitat.

ID: the most familiar white; black front wing tips with at least one black spot on the front wings. *Male:* one black spot on the front wing. *Female:* two black spots on the front wing.

Similar Species: Some Cabbage Whites are nearly unmarked, and can be confused with summer generation *Margined Whites* (p. 73) and *Veined Whites* (p. 72). In turn, some spring generation Margined and Veined Whites have dark forewing spots, but they also have dark hind wing veins, which the Cabbage lacks.

Caterpillar Food Plants: mustard family plants, especially cultivated ones (Brassicaceae).

Habitat & Flight Season: found mostly in or near cropland, gardens or weedy places, but also present in most meadows and open places; multiple broods, up to four in warmer areas; flies throughout the butterfly season.

Large Marble

Euchloe ausonides

Wingspan: 35–55 mm

You may see this species listed as the Dappled Marble, or the Creamy Marblewing, as well as under the scientific name *Euchloe ausonia*. If so, don't let it trouble you because the current nomenclature seems stable and uncontroversial. This butterfly is not particularly drawn to hilltops, and the males spend every available moment of warm, sunny weather flying back and forth through its forest habitat looking for females. If the weather isn't quite warm enough for flight, both sexes will bask in the sun with their wings either opened to the sides, or closed over their backs,

Similar Species

Northern Marble

Green Marble

Desert Marble

allowing a good look at all of their identifying features. Of course, they take time to feed while they are on the wing, preferring white and yellow flowers, which are often the flowers of their caterpillars' host plants. This habit may help the male in his search for an unmated female and getting on with the business of laying eggs. When a male approaches a female, he hovers above her. If she wishes to reject him, she points her abdomen upward and opens her wings part way. This is the rejection posture typical of whites, marbles, and orange-tips. It may physically prevent the male from copulating, or it may simply serve as a display to indicate that the female is not receptive. Either way, it achieves the desired effect.

Also Called: Dappled Marble, Creamy Marblewing; *E. ausonia*.

ID: green and white marbling on the hind underwing covers a roughly equal area; patches of marbling are fairly broad; under magnification, you can always see some white scales in the black cell bar on the upperside of the front wing; cell spot on the front wing is larger on the female than on the male; hind wings may be yellow on the female.

Similar Species: *Northern Marble* (p. 79): has finer marbling. *Green Marble* (p. 81): has sootier marbling. *Desert Marble* (p. 82): has no white scales on the upper front wing cell bar.

Caterpillar Food Plants: mustard family plants (Brassicaceae), especially *Sisymbrium* species and rock cress (*Arabis* species).

Habitat & Flight Season: openings in dry forests; prefers pines; emerges in late spring at low elevations and in early August at higher elevations.

Northern Marble

Euchloe creusa

Wingspan: 35–45 mm

This is a butterfly of the north and of the mountaintops. It was originally discovered at Rock Lake, Alberta, not far from the Continental Divide. And if you're having troubles seeing the streaky as opposed to the blotchy marbling, I apologize. These sorts of features are often easiest to discern with a museum collection. A few dozen Large Marbles in a drawer, all mounted with their undersides up, contrast beautifully with an equal number of Northern Marbles. In the field, however, the distinctions are not so obvious, especially if you are working with a poor view of a single butterfly. This is when you have to use all the tools available to you. First, it is not cheating to look at the

Similar Species

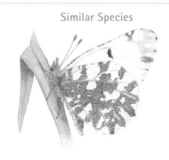

Large Marble

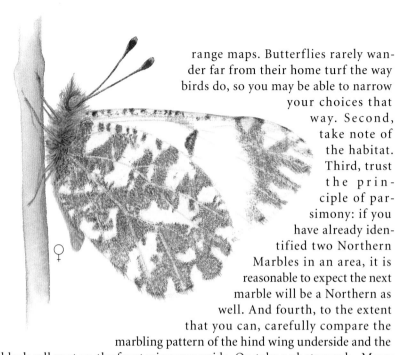

range maps. Butterflies rarely wander far from their home turf the way birds do, so you may be able to narrow your choices that way. Second, take note of the habitat. Third, trust the principle of parsimony: if you have already identified two Northern Marbles in an area, it is reasonable to expect the next marble will be a Northern as well. And fourth, to the extent that you can, carefully compare the marbling pattern of the hind wing underside and the black cell spot on the front wing upperside. Or, take a photograph. Many white butterflies can be notoriously difficult to photograph well, because cameras try to make white things gray, but with digital technology it is easy to put the white back in a white butterfly. And, it's a lot easier to compare a complex wing pattern with the illustrations in the book when there is a photograph handy.

ID: marbling arranged loosely in streaks or bands (squint your eyes if you can't see this); some show a sort of streaking of the marbling, giving the impression of concentric crescents with their concave sides all oriented toward the wing base, but this is not always easy to see (or to imagine); green patches are never outlined in yellow; cell spot on the front wing is larger on the female than on the male.

Similar Species: the Northern Marble has finer patches of marbling than other marbles. *Large Marble* (p. 77): has more yellow.

Caterpillar Food Plants: known to feed only on the lance-leaved draba (Brassicaceae): *Draba cana*).

Habitat & Flight Season: found in openings in dry forests, as well as on alpine tundra; emerges in mid-summer.

Green Marble
Euchloe naina
Wingspan: 35–40 mm

The first Canadian records for this Siberian and Alaskan species were obtained by Ted Pike while he was only 16 years old. He had a freshly issued driver's licence and a used Volkswagen Beetle; his dream was to head off from Calgary and travel the Yukon and collect butterflies alone, camping along the way. In those days (1972) the roads on the way to and within the Yukon were atrocious, and it was a genuine feat to launch an expedition up there, especially for a teenager. Ted's records, as well as those gathered by Felix Sperling and Gerry Hilchie while they were both teenagers as well, gave us our first really detailed impression of the butterflies of the Yukon. The British Columbia records for Green Marbles came more than two decades later, as a result of field work by Norbert Kondla and Jon Pelham in the northwest corner of the province. This species of butterfly is still, by any standard, poorly known in Canada.

ID: more sooty green than white on the hind wing underside; upper wing bases and edges are dark-coloured, often with some greyishness, especially on the hind wing; the upperwing surfaces are somewhat pearly, and the underwing surfaces are a bit silvery; cell spot on the front wing is larger on the female than on the male.

Similar Species: *Large Marble* (p. 77): has less soot-coloured marbling and less rounded wing tips.

Caterpillar Food Plants: mustard family plants (Brassicaceae), but so far there are no records of this caterpillar in B.C.

Habitat & Flight Season: scree and alpine tundra above 1000 metres elevation; emerges in June.

Desert Marble
Euchloe lotta
Wingspan: 40–45 mm

Male Desert Marbles, like many desert butterflies, go to flowery hilltops to find their mates. This tendency, along with other more subtle features, helps to differentiate the Desert from the Pearly Marble (*E. hyantis*), a very close relative. Just as the Desert and Large Marbles resemble each other most closely where they are both found, Desert Marbles are most like their close relatives, the Pearly Marbles, where they occur together in California. It seems likely that this is a sign that evolution has fine-tuned these species pairs to within exacting tolerances of each other in the same evolutionary crucibles. It is also a fairly widespread phenomenon among butterflies, and is especially vexing among the greater fritillaries (*Speyeria* spp.) in the western half of North America. The Desert Marble is also called the Western Marble, and you will still see it listed in some books as a subspecies of the Pearly Marble (*Euchloe hyantis lotta*).

Similar Species

Large Marble

Also Called: Western Marble; *Euchloe hyantis lotta*.

ID: large patches of marbling, with no yellow separating the smaller, pearly white areas on the hind underwing; cell spot on the front upper wing contains no white scales; cell spot on the front wing is larger on the female than on the male.

Similar Species: most like the *Large Marble* (p. 77), but the wings are never yellow, and the cell spot on the upper surface of the front wing has no white scales.

Caterpillar Food Plants: mustard family plants (Brassicaceae), both flowers and leaves.

Habitat & Flight Season: dry grasslands with sage; emerges in April and May.

Sara Orangetip
Anthocharis sara

Wingspan: about 40–45 mm

I have always known this species as the Sara Orangetip, and I'm not at all keen on changing its name. It is a familiar butterfly with a long history in the field guides. Still, there are some who would prefer to call it the Sara's Orangetip and others who refer to it as the Pacific Orangetip or the Western Orangetip. Since the Sara and Stella Orangetips were recognized as distinct species in 1986, it has been tempting to call the remaining subspecies within the original concept of Sara something other than "Sara." This would allow us to distinguish the broad definition of Sara from the narrow definition, but I think this is a mistake because most butterfly enthusiasts do not remember the old ideas, and those who do have no trouble setting them aside. If there is one thing that butterfly

Similar Species

Stella Orangetip

Sara Orangetip

taxonomists seem to take to, it is changes in names. The only other people I have encountered with such a deep fascination with continually renaming things for the sake of renaming them are government bureaucrats and the people who provide them with their logos and letterhead.

This butterfly, by the way, is an amazing creature under ultraviolet light, viewed with the right equipment. The wings are mostly dark and nonreflective, except for the orange wing tips, which reflect UV brightly. Stella Orangetips show the same pattern, and together they are two of the best examples of butterflies whose visible and UV reflectances are strikingly different (see p. 27).

ID: a small white or yellow butterfly (either sex can be yellowish) with orange front wing tips and fine marbling on the hind wing underside. *Male:* no white on the orange tip of the front wing; orange tip is separated from the rest of the wing by a black bar. *Female:* white and black patches on the orange tip; orange tip is not separated by a continuous black bar.

Similar Species: *Stella Orangetip* (p. 85): has lighter-coloured marbling (more yellow scales within the green) on the hind wing underside.

Caterpillar Food Plants: mustard family plants (Brassicaceae), especially rock cress (*Arabis* spp.), feeding on the leaves when young and the flowers when larger.

Habitat & Flight Season: brushy clearings and open hillsides, and in various other dry, open habitats; this is a spring butterfly in the lowlands, emerging as early as March, but an early summer butterfly in cooler and higher places.

Stella Orangetip
Anthocharis stella
Wingspan: about 40–45 mm

In 1986, Hansjürg Geiger and Arthur Shapiro (of Mustard White fame—see pages 18 to 20 for more information about these lepidopterists in the section on butterfly names in the introduction) studied the Orangetips using enzyme electrophoresis, and concluded that the Sara and Stella Orangetips were separate species that do not interbreed to any significant extent. Indeed, in British Columbia, the Stella is a butterfly of the Interior and the Rocky Mountains, while the Sara sticks to the coast. Still, the rest of the butterfly community has been slow to adopt this finding, perhaps because it is based on things that cannot be seen directly. You will therefore find the Stella

Similar Species

Sara Orangetip

Stella Orangetip

Orangetip listed as the Stella Sara Orangetip in many books today (as *Anthocharis sara stella*). You will also see it spelled with an apostrophe (Stella's Orangetip) by those who feel that the mystery woman after whom entomologist and taxonomist W.H. Edwards named this species does not get full recognition for her contribution to lepidoptera science without the possessive form of the name. To my mind, this is just one more way of confusing people, so I ignore it. But perhaps that is confusing as well. It really is difficult knowing how to present the complexity of butterfly nomenclature without oversimplifying or misrepresenting the subject.

ID: a white or yellow butterfly with orange front wing tips. *Male:* has orange tip on an otherwise white wing, separated from the rest of the wing by a black bar. *Female:* orange tip is not separated from the rest of the wing by a black bar; wing may be white or yellow.

Similar Species: compared to the *Sara Orangetip* (p. 83), the marbling on the hind wing underside is more yellowish.

Caterpillar Food Plants: probably feed on rock cress (Brassicaceae: *Arabis* spp.).

Habitat & Flight Season: brushy forest clearings, but also on alpine tundra; flies in April or later in the lowlands, and in early summer at higher elevations.

Sulphurs
(subfamily Coliadinae, genus Colias)

Pink-edged Sulphur

British Columbia is "Sulphur Central," with more species in the genus *Colias* than any other place on earth. Sulphurs are a delightful group of bright, attractive, medium-sized butterflies that spend their time cruising low over the ground, in open areas where they are easy to find, but tricky to approach closely. Most are yellow or orange, though some females of each species are largely greenish white (and look like fast-flying Cabbage Whites). Both sexes of one species, the Nastes Sulphur, are also greenish white with striking pink trim.

Sulphurs almost never spread their wings while perched, so butterfly watchers learn to use the undersides of the hind wings when identifying them. If you think this makes things difficult, you are correct. Catch and release, or collecting specimens, makes the job a bit easier but, to be honest, the sulphurs are just plain difficult even when you have long series of pinned specimens to work with. Humility, patience and caution are the best tools for sulphur identification.

Some male sulphurs have patches on the upperside of the wings (the front wings, hind wings or both). that strongly reflect ultraviolet light. If you have the equipment to see them, these patches can help with tricky identifications.

To identify sulphurs, pay special attention to the following details: 1) What does the cell spot in the middle of the hind wing underside look like? Is it surrounded by a border? Is the border narrow or thick?

Is it uniform in thickness all around the spot, or smeared out like a comet's tail? (When the border is thick, the outer and inner margins are generally darker than the intervening area, and the border is often said to be "doubled.") 2) Does the cell spot have a smaller "satellite spot" adjacent to it? (The main spot is traditionally called the "discocellular spot," a term that for me brings images of bell-bottom pants and portable phones to mind.) 3) Are there one to three small, dark, submarginal spots parallel to the outside margin of each wing, on the underside? 4) What are the dark borders like on the upper surfaces of the wings? (If the sulphur is perched, you may be able to see the outline of these borders through the translucent wings.) 5) What colour is the butterfly, generally, as it flies past? (You may have to look closely to see orange.)

Faced with an unidentified sulphur, your goal should be to find your way quickly to the right portion of the sulphur clan rather than wallowing in the genus as a whole. To this end, let's divide the British Columbia sulphurs into four groups. First, there are the two most common species, which are also their own closest relatives: the Clouded and Orange Sulphurs. Get to know these two well because they will make up the bulk of your sightings.

Second, there are the species with yellow males that do not reflect UV: the Giant, Pink-edged, Pelidne and Palaeno Sulphurs. The Western Sulphur is treated here as part of this latter group, but is more closely related to the Alexandra-group sulphurs.

Male Clouded Sulphur, from a dandelion patch outside the publisher's office

The species in the Alexandra group—the third subgroup—have males with orange on the wings (the Queen Alexandra's is yellow without orange) that reflect UV: the Queen Alexandra's, Christina, Mead's, Hecla and Canada Sulphurs. The Nastes Sulphur belongs here, too, technically, but the males of this species are so different in appearance that I think of them on their own, as a fourth category of B.C. sulphur.

This system will allow you to identify males with relative ease. Females will be trickier, especially white females. I am confident, however, that by examining the butterfly, and taking note of the habitat, flight season, geographic range and the presence of identifiable males at the same site at the same time, you will be able to put a name on most of the sulphurs you see.

Clouded Sulphur
Colias philodice
Wingspan: 35–55 mm

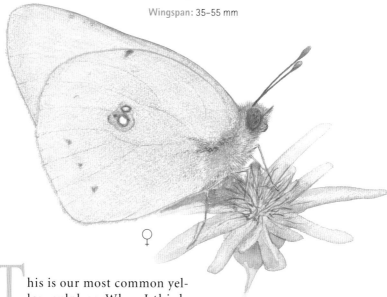

♀

This is our most common yellow sulphur. When I think of Clouded Sulphurs, three images come quickly to mind. First, I see them cruising over a hay field, dodging left and right and occasionally gathering together into a short-lived swarm around a particular female, ascending higher and higher into the sky. Then I picture a male sipping nectar at a blossom while I try to confirm his underside field marks. And finally, I see in my mind's eye a group of 20 to 100 of these lovely yellow butterflies crowded alongside a roadside mud puddle in mid-summer. Spooked into flight, they fill the air with shimmering yellow, made all the more glorious by the Greenish Blues that are sipping for minerals alongside them.

Clouded Sulphurs are very familiar butterflies, and may well embody the origin of the word butterfly as well because they are both yellow and airborne. Etymologists (as opposed to entomologists) disagree strongly, however, as to whether this explanation is genuine, and we will probably never know where the

Similar Species

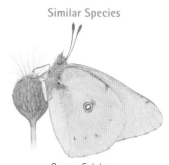

Orange Sulphur

Clouded Sulphur

word truly comes from. The scientific name *philodice* is also slippery, and may mean either "lover of the seasons" or "lover of lawsuits." Considering how many arguments I have heard regarding sulphur identification, I vote for the latter. This species is also known as the Common Sulphur and the Yellow Sulphur.

white morph ♀

yellow morph ♀

ID: most have three dark, submarginal spots on the underside of each wing, and a silvery cell spot with a thick red border and a satellite spot. *Male:* does not reflect UV; upper dark wing borders solid and relatively wide. *Female:* light spots within dark forewing border; ground colour may be yellow or white; white females are more common in the north, and yellow in southern B.C.

Similar Species: *Orange Sulphur* (p. 91): at least part of the upper wing is orange, except for white females, which are on average larger, with bigger, darker submarginal spots, and wider dark wing borders.

Caterpillar Food Plants: legumes (Fabaceae), leading to abundance in areas with lots of clover or alfalfa.

Habitat & Flight Season: open areas in a variety of habitats, especially disturbed areas and agricultural land; flies from May until October at lower elevations, but only in mid-summer at higher elevations; in warmer places, where they produce three or four broods, these are typically the last butterflies seen in the fall.

Orange Sulphur
Colias eurytheme
Wingspan: 35–60 mm

♂

Also called the "Alfalfa Butterfly," this species has long been thought to interbreed with the Clouded Sulphur. For a time, lepidopterists believed that they were both a part of one widespread, variable species. This is no longer thought to be the case because female Orange Sulphurs will not mate with males that do not reflect UV. Orange Sulphurs also require their own species' courtship perfumes before mating. However, it turns out that they don't quite understand these rules until they have been out of the pupa for at least an hour, and in the life of a butterfly, a lot can happen in that short time. Thus, some hybrids do occasionally arise. To make the story even more complex, male Clouded Sulphurs that mate with female Orange Sulphurs produce sterile female young, while the opposite cross is fertile. Oddly enough, female butterflies get their single X chromosome from

Similar Species

Clouded Sulphur   Christina's Sulphur

Orange Sulphur

their fathers, just as people get their single Y chromosome from their fathers, and it is the genes on the X chromosome which determine the type of male a female sulphur will be attracted to. So, even with limited interbreeding, Orange-Clouded Sulphur hybrid females return to mate with their parent species, the Orange Sulphur, maintaining the distinction between these two species of butterflies.

Orange Sulphurs spend the winter as caterpillars, as do all other temperate sulphurs. They cannot, however, survive the cold in British Columbia, so this species has to recolonize the province each year. The first Orange Sulphurs of spring are American immigrants, followed by a generation of small butterflies that develops here, and then a generation of normal-sized butterflies that persists into the fall. Most probably die with the onset of cold weather, but there is some evidence that Orange Sulphurs may move to the south as the days become shorter in October.

ID: upper side is mostly orange, at least at the wing bases, except in white females; submarginal spots are always present on the underside; cell spot has thick pink border and often a satellite; first generation butterflies each year are smaller than average, with narrower dark wing borders. *Male:* orange on upperwing reflects UV strongly; black wing borders are solid and relatively wide. *Female:* usually orange, but often white; yellow spots on the wing borders of orange females.

Similar Species: *Clouded Sulphur* (p. 89): male may have some orange on the underside of the forewing, but not on the upper side, and does not reflect UV; female has no orange on the upperwing, smaller submarginal spots and narrower wing borders. See also *Christina's Sulphur*. All other orange sulphurs are darker on the underside.

Caterpillar Food Plants: all types of legumes (Fabaceae).

Habitat & Flight Season: open areas and fields; flies from May to October, and generally shows three broods each year.

Giant Sulphur
Colias gigantea
Wingspan: 40–60 mm

Wouldn't it be simple if the Giant Sulphur really was bigger than all the rest? Sigh. Nothing about Sulphur identification is simple, it seems. Fortunately, Giant Sulphurs are usually found in their own unique habitats, near willows in wet meadows. The very similar and closely related Pink-edged Sulphur is much more likely to be found in open, dry forests, and it also emerges later in the season than does the Giant. As you can see from the measurements, however, large Pink-edged Sulphurs are easily as large as small Giants, so one really does have to look at them carefully for a confident identification.

In British Columbia, this species is found in three isolated regions: the northeast, the extreme northwest and the Cariboo Region of the northern Interior. Those in the extreme northwest belong to a separate subspecies (*Colias gigantea gigantea*) than the others (*C. g. mayi*), and the former race of Giant Sulphur will show up in some books under the name *Colias scudderii* (the Scudder's or Scudder's Willow Sulphur). The "Scudder" in question, by the way, is not the well-known, living B.C. entomologist Geoff Scudder, but the long-dead American lepidopterist Samuel Scudder, who wrote many of the first important publications on North American butterflies. There are some butterflies named for British Columbians, but this isn't one of them.

Similar Species

Pink-edged Sulphur

ID: a bright yellow sulphur with relatively little black dusting on the underwing; cell spot is large, with a brown rim and often a satellite. *Male:* inner margin of the solid, dark forewing border is roughly straight along its middle region. *Female:* dark wing borders almost nonexistent; some females are white.

Similar Species: *Pink-edged Sulphur* (p. 94): forewing has a more rounded hind angle, which is not visible in the illustration above, or through binoculars.

Caterpillar Food Plants: willows (Salicaceae: *Salix* spp.), unlike all of our other sulphurs.

Habitat & Flight Season: wet places with willow shrubs; flies in mid-summer.

Pink-edged Sulphur

Colias interior
Wingspan: 40–50 mm

The pink wing fringes of the Pink-edged Sulphur are indeed brighter and more prominent than those of any of our other species, but you'd be foolish to use them as the only means of recognizing this species. Older individuals lose the fringe, and just about every other sulphur in the province has some sort of "pink-edge," as well. The scientific name, *Colias interior,* works better as a mnemonic, if only because this is indeed a butterfly of the Interior (and the northeast) of British Columbia. It is a widespread species in interior North America (where the actual meaning of the name originates), and is probably the next most familiar sulphur after the Clouded and the Orange.

Like two of its close relatives, this species has a specialized diet of blueberries, an unusual food plant for sulphurs. This alone is evidence that the Pelidne, Palaeno and Pink-edged Sulphurs (the "three Ps") are not only related, but also similar in appearance. However, it is the Giant Sulphur that most people confuse with the Pink-edged.

Similar Species

Giant Sulphur

ID: upper surface is a rich yellow; underside has a little dark dusting, a silver cell spot with a pink-red rim and sometimes a red satellite and no submarginal spots. *Male:* the inner margin of the dark forewing border is curved in its middle region. *Female:* the forewing tip is usually darkened; white females are rare.

Similar Species: *Giant Sulphur* (p. 93): has a less evenly rounded hind angle on the forewing (not visible in the illustration above, or through binoculars).

Caterpillar Food Plants: blueberry leaves (Ericaceae: *Vaccinium* spp.).

Habitat & Flight Season: openings in forests; flies from June through early September.

Pelidne Sulphur

Colias pelidne

Wingspan: 35–45 mm

P ell-IDD-nee: it's not hard to pronounce. The name is worth learning, too, because this species is sometimes called the Labrador Sulphur, a name also assigned to the Nastes Sulphur. If you're wondering what Labrador has to do with butterfly names in British Columbia, I'm with you. Still, the Pelidne Sulphur is a rare denizen of the Rockies that you are not likely to encounter very often, let alone ponder in a nomenclatorial sense.

The geographic range of this species is unusual in that it encompasses three separate regions: the eastern Arctic, the western Arctic and the Rocky Mountains. It is always difficult to determine whether populations that never meet really belong to the same species, so we will probably continue to see disagreements over this sulphur's status. In the past, many butterfly scientists considered it to be the same species as the Pink-edged Sulphur, an idea that may or may not resurface in the future. But on the Alberta side of the Continental Divide, where it flies alongside the Pink-edged, it seems unlikely that they are all one species. If you want to see this species for yourself, a trip to Alberta is probably the easiest way to find one.

Similar Species

Pink-edged Sulphur

Palaeno Sulphur

ID: pale yellow with dark wing bases on the upper side and heavily dusted greenish on the under side; cell spot has a red rim and is often dusted with red. *Male:* solid wing borders. *Female:* wing borders are absent, or narrow and scalloped; white females are common.

Similar Species: *Pink-edged Sulphur* (p. 94): much less dark dusting. *Palaeno Sulphur* (p. 96): more rounded wings and no red rim on the cell spot (but the rim can be absent in Pelidne Sulphurs, too).

Caterpillar Food Plants: blueberry leaves (Ericaceae: *Vaccinium* spp.).

Habitat & Flight Season: alpine tundra and meadows; flies from late June to early August.

Palaeno Sulphur
Colias palaeno
Wingspan: 40–45 mm

This is a butterfly of the northern quarter of the province. Like the Nastes Sulphur, it carries with it a bit of a wilderness mystique. The subspecies we have in British Columbia (*Colias palaeno chippewa*) is thought by some to be a species in its own right, and is often listed as such, under the name "Chippewa Sulphur." Part of the difficulty in knowing whether it should be considered a species or a subspecies stems from the fact that this is another sulphur that can be found in all the northern areas of the globe. The Eurasian populations (where the name Palaeno originated) are isolated from the North American ones, making the usual criteria for species recognition difficult or impossible to apply. In other words, we don't know if they could interbreed if they were to meet in the wild (although there is some evidence that the two types of sulphurs are separate in eastern Siberia). For the moment, however, I am taking the conservative route and continuing to call these butterflies by the older name, Palaeno Sulphurs, if only just to more clearly connect with the other butterfly books in print.

Similar Species

Pelidne Sulphur

Pink-edged Sulphur

Also Called: Chippewa Sulphur; *C. chippewa*.

ID: another small yellow sulphur with extensive dark dusting; cell spot is small, with a very faint, dark rim, or none at all. *Male:* relatively wide, solid, dark wing borders. *Female:* wing borders may be solid black, or contain pale spots; white females are more common than yellow.

Similar Species: *Pelidne Sulphur* (p. 95) from the Rockies, or a *Pink-edged Sulphur* (p. 94), but without a heavy rim on the cell spot.

Caterpillar Food Plants: blueberry leaves (Ericaceae: *Vaccinium* spp.).

Habitat & Flight Season: alpine tundra and forest clearings; flies from mid-June to early August.

Western Sulphur
Colias occidentalis
Wingspan: 45–55 mm

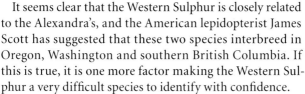

♂

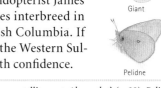

Similar Species

Pink-edged

Giant

Pelidne

The Western Sulphur is a lot like the Sasquatch—how often you see it is directly related to how strongly you believe it actually exists. And just as bears, tree stumps and ordinary people can be mistaken for giant furry primates, the Western Sulphur has a few look-alikes in British Columbia as well, most notably the Clouded, Alexandra's, Giant and Pink-edged Sulphurs. Most Clouded Sulphurs will have a double rim around the cell spot, while most Alexandra's Sulphurs will be more greenish than yellow on the underside of the hind wing and the males will reflect UV light. Pink-edged and Giant Sulphurs show a lighter shade of yellow on the hind wing underside, especially near the wing base.

It seems clear that the Western Sulphur is closely related to the Alexandra's, and the American lepidopterist James Scott has suggested that these two species interbreed in Oregon, Washington and southern British Columbia. If this is true, it is one more factor making the Western Sulphur a very difficult species to identify with confidence.

ID: upper side is a yellow ground colour; hind wing underside rich, dark yellow with relatively little black dusting and no submarginal spots (or faint, tiny ones); large cell spot with only a single, narrow red rim. *Male:* solid, dark wing borders; wings do not reflect UV. *Female:* varies from very pale yellow to almost as dark as the male; faint dark border on forewing (with yellow spots in it), dark tip on hind wing.

Similar Species: *Clouded Sulphur* (p. 89): thick cell spot rims, often a satellite and submarginal spots. *Pink-edged Sulphur* (p. 94) and *Giant Sulphur* (p. 93): even more like Westerns, but lack the dark yellow hind wing underside, and often show a satellite spot. *Alexandra's* (p. 98), *Pelidne* (p. 95) and *Palaeno Sulphurs* (p. 96): all have more heavily dusted hind wing undersides.

Caterpillar Food Plants: bean family plants (Fabaceae), probably vetch and lupine in B.C. (*Astragalus* spp. and *Lupinus* spp.), but possibly clover (*Trifolium* spp.).

Habitat & Flight Season: grassy, relatively open, dry areas with conifers interspersed or nearby; flies from May through August.

Alexandra's Sulphur
Colias alexandra
Wingspan: 40–60 mm

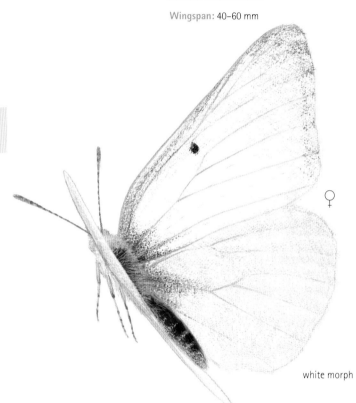

white morph

Similar Species

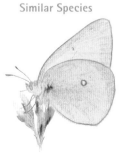

Western Sulphur

The Alexandra's Sulphur is commonly referred to as the Queen Alexandra's Sulphur, but I have read arguments that it was either named after an Alexandra who was not a queen, or after Queen Alexandra of Denmark before she was crowned. It therefore makes sense to call this species the Alexandra's Sulphur, and thereby shorten the name without causing too much unnecessary confusion. Like all sulphurs, this one comes with a bit of built-in confusion; it was once believed to include the Christina's Sulphur as a subspecies (in other words, a local

Alexandra's Sulphur

geographic race). Two recent studies have refuted that notion, and it is now fashionable to think of these closely related sulphurs separately: the yellow Alexandra's and the orange Christina's. Will this arrangement last? At least in the foreseeable future I believe it will, if only because the process of classification these days leans toward "splitting" species into their smaller subgroups, rather than "lumping" them together.

ID: light yellow on the upper side, with a greenish yellow hind wing underside and a narrow rim on the cell spot. *Male:* reflects UV, especially from a patch on the top of the hind wings (which is a darker shade of yellow on fresh individuals). *Female:* either white (all of the northern females in B.C.), or very pale yellow; only shows a faint dark wing margin.

Similar Species: *Western Sulphur* (p. 97): male is a darker yellow and does not reflect UV; female has a darker hind wing underside and a darker cell spot rim.

Caterpillar Food Plants: bean family plants, especially milk-vetch (Fabaceae: *Astragalus bisulcatus*).

Habitat & Flight Season: grasslands and open, dry forests; flies from late May to August; earlier at lower elevations.

Christina's Sulphur

Colias christina

Wingspan: 35–55 mm

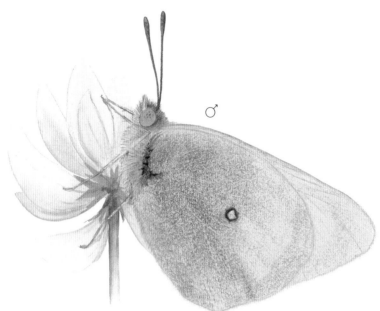

The Christina's Sulphur, like just about all other sulphurs, has a long history of being confused with other species. It has been considered a race of both the Alexandra's Sulphur and the Western Sulphur, and is still not recognized as a separate species by most American butterfly experts. Canadian experts seem to agree that it is worthy of its own species name, and, because most Christina's Sulphurs live in Canada, I trust the Canadian opinion on this matter.

This is a butterfly of British Columbia's northeast corner, and is much more common east of the Continental Divide, in

---

Similar Species

Orange Sulphur

Mead's Sulphur

Canada Sulphur

Hecla Sulphur

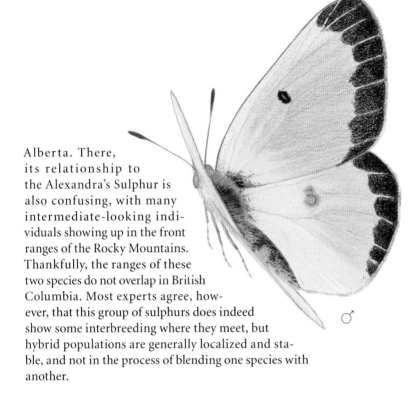

Alberta. There, its relationship to the Alexandra's Sulphur is also confusing, with many intermediate-looking individuals showing up in the front ranges of the Rocky Mountains. Thankfully, the ranges of these two species do not overlap in British Columbia. Most experts agree, however, that this group of sulphurs does indeed show some interbreeding where they meet, but hybrid populations are generally localized and stable, and not in the process of blending one species with another.

ID: has some orange on the upper wings, except on white females; otherwise very similar to Alexandra's Sulphur, with an olive or olive-yellow hind wing underside; rarely with faint submarginal spots. *Male:* reflects UV from patches on both front and hind wings. *Female:* most are orange, some are white.

Similar Species: *Orange Sulphur* (p. 91): orange located at the base of the wings, whereas on the Christina's, it is located along the outer half of the wings. *Mead's, Hecla* and *Canada Sulphurs* (pp. 102–05): all are darker, with more extensive orange colouration, and are generally smaller.

Caterpillar Food Plants: bean family plants, such as sweet vetch and buffalo bean (Fabaceae: *Hedysarum sulfurascens* and *Thermopsis rhombifolia*).

Habitat & Flight Season: open areas in relatively dry, coniferous forests; flies from June to September, but is most common in July.

Mead's Sulphur
Colias meadii
Wingspan: 35–45 mm

This is a rare species in British Columbia, and another butterfly that is easier to find by driving across the Continental Divide, into Alberta. In the cool environments where they live, it is not surprising that these butterflies show special adaptations. The caterpillars probably overwinter twice in British Columbia, taking two years to reach adulthood. The adults are also strongly adapted to cold places.

Similar Species

Orange Sulphur

Christina's Sulphur

Hecla Sulphur

The dark colour of their wing bases allows them to warm up more efficiently when the sun shines, and in turn this allows them to make the best of their brief lives. These butterflies live less than a week as adults, and strive to mate and lay eggs in this time, weather permitting. Fortunately, because they can achieve these objectives in their first day of life as a butterfly, all they need is one good sunny mountain afternoon to ensure that the next generation will live on. Wind doesn't seem to bother them, and, in fact, Mead's Sulphurs live in some of the windiest habitats in the province. Unlike most North American sulphurs, Mead's Sulphur was not named for a woman. Theodore Mead was an American butterfly collector, and as far as I know this is the only species named in his honour. You might encounter fish named *"meadi"* but these names honour Giles Mead, an ichthyologist at Harvard.

More Similar Species

Canada Sulphur

Nastes Sulphur

ID: a sulphur of the mountains, with dark orange upperwings; underside is greenish yellow or greenish orange, with wing borders vaguely paler than the rest of the wing; a thin, red rim around the plain cell spot; no submarginal spots. *Male:* wide black wing borders; reflect both violet and UV from both the front and hind wings, making special viewing equipment unnecessary if the butterfly is fresh and in the hand. *Female:* yellow spots in the wide, but ill-defined, wing borders; white females are rare.

Similar Species: *Orange Sulphur* (p. 91) and *Christina's Sulphur* (p. 100): both have less extensive orange colour on the upperwings. *Hecla Sulphur* (p. 104) and *Canada Sulphur* (p. 105): both have a larger cell spot that is often smeared outward. White females could be confused with the *Nastes Sulphur* (p. 106), but Nastes are darker overall.

Caterpillar Food Plants: bean family plants, such as clovers that can grow in the alpine zone (Fabaceae: *Trifolium* spp.).

Habitat & Flight Season: alpine meadows in the Rocky Mountains; flies in midsummer only.

Hecla Sulphur
Colias hecla
Wingspan: 35–45 mm

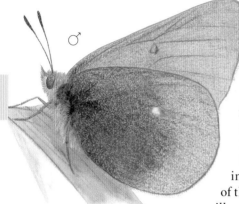

"Hecla" is not a person's name. It is apparently an Icelandic word for "hill." Hecla Sulphurs are found all the way from Alaska to Greenland, across the Arctic (as well as in northern Europe and Asia), and in British Columbia they are found only in the extreme northwest corner of the province. In some books, you will see them referred to as Greenland Sulphurs because they are the only species of sulphur found in Greenland. That's nice for Greenland, but it isn't a good enough reason for us to call them Greenland Sulphurs here in Canada, in my opinion. In other books, you will see this species' range extending into much of northern British Columbia and the northern Rocky Mountains of Alberta. This is because the Canada Sulphur was once considered part of the same species as the Hecla Sulphur. The separation of the two is based on the work of Clifford Ferris, an American lepidopterist who is currently the best-respected sulphur expert on this continent. He found that he could recognize Canada and Hecla Sulphurs as distinct species in Alaska, not far from where they coexist in northern British Columbia.

Similar Species

Mead's Sulphur

ID: a dark orange sulphur with heavily dark-dusted wing bases on the greenish underside; the pink rim of the cell spot is smeared toward the outer edge of the wings, resembling a comet's tail. *Male:* solid dark wing borders, both wings reflect UV. *Female:* wide, dark wing borders with pale spots; white females occur regularly.

Similar Species: *Canada Sulphur* (p. 105): lighter orange, with narrower dark wing borders. *Mead's Sulphur* (p. 102): has a smaller cell spot that is never smeared like a comet.

Caterpillar Food Plants: bean family plants, such as clovers and alpine milk vetch (Fabaceae: *Trifolium* spp. and *Astragalus alpinus*).

Habitat & Flight Season: wetter parts of the alpine zone; flies in the warmest part of summer only.

Canada Sulphur
Colias canadensis
Wingspan: 35–50 mm

We are lucky that any given locality in British Columbia will have only one or two species of orange-coloured sulphurs. A careful butterfly enthusiast will examine each individual with an open mind, but the fact remains that you can expect certain species in certain places, and this will save you a great deal of confusion and distress.

The "type locality" for the Canada Sulphur species is mile 209 on the Alaska Highway. I doubt this will ever make it into the *Milepost,* the guide that most travelers use when making the summer trek up the highway, but it is still worth noting. And what is a type locality? Well, let's start with what it is not—it is not the place where the most typical members of a given species are found. It is, however, the place where the "type specimen," or first-described specimen, of the species came from. The name *Colias canadensis* will always refer to the type specimen and all other individuals in the same species. Thus, if in the future someone demonstrates that some of our Canada Sulphurs are not Canada Sulphurs after all, the "real" ones will be those that live at the type locality, along with others of the same persuasion.

Similar Species

Mead's Sulphur

ID: a light-coloured orange sulphur, with a greenish underwing and a faintly smeared, red cell spot rim (although sometimes not smeared at all); no submarginal spots. *Male:* dark wing borders relatively narrow; upperwings reflect UV. *Female:* most are white, with wide, but faint wing borders.

Similar Species: *Hecla Sulphur* (p. 104): darker, with wider dark wing borders. *Mead's Sulphur* (p. 102): has a smaller cell spot.

Caterpillar Food Plants: bean family plants (Fabaceae).

Habitat & Flight Season: open, shrubby areas in the northern forests; flies from June through August.

Nastes Sulphur
Colias nastes
Wingspan: 30–45 mm

♀

Labrador Sulphur is the name that most American field guides now use for this species. But as with the Greenland Sulphur, I can't help thinking it is silly to name a butterfly after some far-flung portion of its range. I also find that I get Greenland and Labrador mixed up quite easily when trying to remember

Similar Species

Hecla Sulphur

Canada Sulphur

Mead's Sulphur

which sulphur is which, while Hecla and Nastes stay straight in my mind. I also dislike the name Arctic Sulphur, which is sometimes used for the Nastes Sulphur, since the Arctic, like Labrador, lies somewhere outside the province of British Columbia. Mind you, most of the time, I dislike English names that simply recast the scientific name as an English equivalent, but in this case the word "Nastes" (pronounced "NASS-teez") is simple, easy to remember and neutral in meaning. In fact, lepidopterists are unsure whether it means "inhabitant," or "firm," neither of which evokes much imagery for me. If you disagree with me, that's fine, too, since butterfly people are still nowhere near standardizing the names for our North American species, despite what some claim.

South of the Canadian border, the Nastes Sulphur is a rare species indeed, and it has a mystique about it similar to that of the gyrfalcon and the boreal owl among birders. Once you've seen a Nastes Sulphur, you'll have a story of high adventure, especially if you take your butterfly interest on the road to the south. Like hunters around a campfire, butterfly people love telling stories of the chase.

The dark scaling on the underside of this butterfly's wings helps warm the body in sunlight. This has been studied in great detail by Ward Watt, a professor at Stanford University, who has done much of his work at the Rocky Mountain Biological Laboratory in Colorado. My friend Felix Sperling had a summer job at the RMBL when he was an undergraduate, and got to know the Watt team. They were in awe of Felix's familiarity with the Nastes Sulphur, a species they had never encountered themselves. As a result, they coined the nickname "Nasty Felix," which he quite enjoyed at the time. Felix Sperling is anything but nasty, by the way, and his patient, firm, and gentle way with students and colleagues is a continual inspiration to me.

ID: a greenish sulphur (up close, it is white with extensive black dusting); dark wing bases; light spots within the dark wing borders; cell spot is silvery and has a smeared pink margin. *Male:* does not reflect UV.

Similar Species: white female *Hecla Sulphur* (p. 104) and *Canada Sulphur* (p. 105): have a bright orange cell spot of the upper surface of the hind wing. *Mead's Sulphur* (p. 102): white female is generally not as darkly dusted.

Caterpillar Food Plants: bean family plants, including alpine milk vetch (Fabaceae: *Astragalus alpinus*).

Habitat & Flight Season: open, windy alpine tundra; flies in mid-summer only.

Gossamer-Winged Butterflies
(Family Lycaenidae)

Introduction to the Gossamer-Winged Butterflies

To recognize a gossamer-winged butterfly, check for the look on its face. Typically, it has big, black eyes surrounded by a white fringe of scales, giving it a characteristic expression that is, frankly, cute. It is also quite literally a matter of family resemblance, since all members of the family Lycaenidae share this "look." Lepidopterists have expressed this in more objective terms, noting that the face is taller than the space between the eyes. Most gossamer-wings are small, in the skipper size range, which also helps with recognition.

Among members of this family, the forelegs of the males are small and not used for walking. The caterpillars are slug-like, in the sense that they are relatively short and fat, and flattened top to bottom; like tiny deflated footballs. Once you see a gossamer wing caterpillar, they become easy to recognize. All of them, except the coppers, possess a gland that produces a honey-like substance that is offered to ants in return for protection from enemies. Some are so closely associated with ants that they cannot survive without them. As well, it is typical for caterpillars of this group to show a variety of colour forms, some green, some brown, some even pink.

The division of the family into coppers, hairstreaks and blues, dates back to the work of two Austrian monks in 1776. In the words of lepidopterist W.J. Holland, "it is therefore as old as the American Republic, and has outlasted almost every government on the globe" (*The Butterfly Book*, p. 222). However, lest you think this is a done deal, many lepidopterists also include the metalmarks within this family, and the pros and cons of this arrangement are still under debate.

Gray Copper (opposite), the cute face of a Spring Azure (above)

Coppers
(Subfamily Lycaeninae)

The coppers are marvelous butterflies, and they are often overlooked. In many, but not all, species the males exhibit iridescent orange or coppery colours, and this is where the coppers get their name. Not

Introduction to the Coppers

Two lovely male Purplish Coppers

all coppers are copper-coloured, but the group still has a look once you get to know them. In general, these are small butterflies that stay close to the ground and spend a fair amount of time chasing each other, other butterflies and even birds and airplanes.

In British Columbia, one member of the copper group is actually blue, and none of our coppers show hind wing tails. This should help you to recognize most coppers in contrast to the blues or hairstreaks, along with the fact that most coppers are faster fliers than blues, but slower than hairstreaks (and fly less frequently than either). Technically, the subfamily is distinguished by details of the structures of the sex organs. As well, males lack androconial wing scales (for the release of courtship pheromones), and the larvae lack honey glands. In general, the way to recognize a copper is to get to know the local species. It's much easier to learn nine butterflies (well, 18 if you count both females and males) than it is to spot the features that indicate why all coppers are believed to have descended from an ancestor that was neither a blue, nor a hairstreak.

American Copper
Lycaena phlaeas
Wingspan: 25–35 mm

♂

This pretty little butterfly is found in both eastern and western North America, but not in between. The eastern race is much more brightly coloured than are the members of the four British Columbian subspecies, but they are, nonetheless, all closely related. Interestingly, some authorities think that the eastern populations may have been introduced from Europe early in the colonization of North America. Across the Atlantic, this species is called the Small Copper, to distinguish it from the Large Copper, *Lycaena dispar*. The Small Copper is certainly more similar to the eastern American Copper than to any of the western races, but some feel that

Similar Species

Lustrous Copper

Mariposa Copper

Dorcas Copper

the two are not identical. We may never know for certain how this species got here, or why the eastern and western races differ as much as they do. Still, even high in the mountains, I can't help being reminded of the classic butterfly books of both Britain and the eastern United States every time I see an American Copper. The look of this butterfly is so distinctive, and so tightly associated with the great lepidopterists in history, that the connection is impossible to ignore.

More Similar Species

Purplish Copper

Lilac-bordered Copper

Bronze Copper

Also Called: Small Copper.

ID: the orange forewing contrasts strongly with the grey hind wing on the topside and underside; some iridescent copper colour on the forewing topside. The orange band on the hind wing is wide on the topside, but narrow on the underside. *Male:* forewing bright coppery on the topside. *Female:* less coppery, but otherwise similar.

Similar Species: *Lustrous Copper* (p. 114): also iridescent, but has black spots on the outside edge of the underside of the hind wing. *Mariposa, Dorcas, Purplish, Lilac-bordered* (pp. 122–28): and *Bronze Coppers* (p. 116): females can look something like American Coppers, but have distinctive undersides.

Caterpillar Food Plants: various sorrel and dock species (Polygonaceae: *Rumex* spp. and *Oxyria digyna*).

Habitat & Flight Season: high in the mountains in clearings and tundra; adults fly from mid-July to mid-August, generally.

Lustrous Copper
Lycaena cupreus
Wingspan: 25–30 mm

In British Columbia, the Lustrous Copper is a high-altitude butterfly, like the American Copper. But, unlike the American, it is truly a western creature and does not live at low altitudes elsewhere. The two species probably both overwinter as eggs or young larvae, and although the American is a stunning butterfly, my vote would go to the Lustrous for pure, well, lustre! The Ruddy Copper (which does not live in British Columbia) beats the Lustrous for sheer iridescent orange, but among the coppery coppers, the Lustrous reigns supreme.

In some books, you will see this species listed as *Lycaena snowi*, named for the former president of the University of Kansas (and the namesake of the Snow Entomological Museum at that institution).

Similar Species

American Copper

The name *Lycaena cupreus snowi* refers to the northern race of the Lustrous Copper, with the southern race living in California, Nevada, Oregon and Utah. Another name that pops up is *henryae*, which, because of the *–ae* ending, is clearly named for a woman (patronyms end in *–i*, as in *snowi*). The woman who collected the first specimens of this subspecies was Jospehine de Henry, but it is still not clear whether this butterfly differs from Professor Snow's, or whether either is sufficiently different from the southern Lustrous Coppers to be considered a separate species.

Also Called: *Lycaena cuprea* or *Lycaena snowi*.

ID: a strongly iridescent, copper-coloured butterfly, with a very faint orange band on the hind wing; wide black margins on the upper surfaces of the wings; male is brighter orange than the female, with smaller dark spots on the underwings.

Similar Species: the only other copper found above timberline in B.C. is the *American Copper* (p. 112), which has a more prominent orange band on the bottom of the hind wing; hind wing underside spots are smaller than those on the forewing; no dark spots along the outside edge of the hind wing.

Caterpillar Food Plants: various sorrel and dock species (Polygonaceae: *Rumex* spp. and *Oxyria digyna*).

Habitat & Flight Season: alpine meadows and tundra, open forests high in the mountains; flies in late July and August.

Bronze Copper
Lycaena hyllus
Wingspan: 35–40 mm

East of British Columbia, this is a common and well-known butterfly, found in almost all grassy and reedy areas near water. The adults perch in the open and are easy to spot in flight. Their grey underwings make them appear light-coloured, and they fly relatively slowly for coppers. Because it is fairly easy to identify, it serves as an easy introduction to the coppers for many people.

Similar Species

Lustrous Copper   American Copper   Purplish Copper

Bronze Copper

Hyllolycaena is now considered a subgenus, within the broader genus *Lycaena*. For this reason, it is OK to place *Hyllolycaena* within parentheses, as in *Lycaena (Hyllolycaena) hyllus*. This does not mean that *Hyllolycaena* is simply another name for *Lycaena*. Rather, it is a subgroup. To show that two names mean the same thing in a particular context, you use an "=" sign. When you see *Hyllolycaena* used as the genus name in place of *Lycaena*, you know you are looking at the work of a "splitter"—someone who likes to emphasize differences. When you see it as a subgenus (or not at all), you are witnessing the "lumper" in action—someone who prefers to emphasize similarities. And if this seems obtuse, consider how something could possibly be both "bronze" and "copper" at the same time.

Also Called: *Hyllolycaena hyllus* or *Lycaena thoe*.

ID: a wide orange band on the hind wing characterizes this butterfly on both the upper and underwing surfaces. On the under side, the forewings are orange while the hind wings are grey. *Males:* brown on the topside with a purplish iridescence and a wide orange hind wing band. *Females:* more orange on the upper forewing.

Similar Species: *Purplish Coppers* (p. 124): male is similar but has fewer dark spots above, a slightly wider orange band, and is slightly larger. *Lustrous Copper* and *American Copper* (pp. 112–14): female is similar but neither has the wide orange band on the hind wing underside.

Caterpillar Food Plants: various sorts of dock (Polygonaceae: *Rumex* spp.).

Habitat & Flight Season: generally a butterfly of open wetlands, but in B.C., it is found in a single small bog area in the northeast; flies in August.

Gray Copper
Lycaena dione

Wingspan: 35–40 mm

British Columbia lepidopterists Crispin Guppy and Jon Shepard call this the most endangered butterfly in British Columbia. The marsh in which it breeds, near Cranbrook, is threatened by park development, road construction and even by management of the marsh for waterfowl. It is often the case that conservation for one sort of animal or plant comes into conflict with conservation for another, and this could become a textbook example. Stories of this sort underline the difference between "conservation" on the one hand and "preservation" on the other. Leaving nature to its own devices, on the assumption that natural things deserve to continue to exist for their own sake, is preservation. Ensuring that particular species, or ecosystems, continue to exist for the benefit of people is conservation (at least in the strict definition of the word, although it

is used in a broader sense most of the time). Thus, the preservation of the Gray Copper is in conflict with the conservation of waterfowl and outdoor recreation opportunities near Cranbrook. Ironically, on the other side of the Rockies, in Alberta, this butterfly can be found in vacant lots and along weedy railway lines, where it is not at all endangered, let alone tied to natural settings.

Also Called: Dione Copper; *Chalceria dione*.

ID: our largest copper; mostly grey on the topside, with grey undersides on both wings; two relatively wide orange bands on the topside and underside of the hind wing. *Male:* grey; forewing has small black spots at the top and a small amount of orange on the outer hind corner; orange hind wing band. *Female:* more orange on the forewing.

Similar Species: a distinctive butterfly.

Caterpillar Food Plants: water smartweed, but possibly various dock species as well (Polygonaceae: *Polygonum amphibium* and *Rumex* spp.).

Habitat & Flight Season: found at one location in B.C., at a marsh near Cranbrook; elsewhere, found in either wet areas or dry weedy areas, in association with their host plants; flies from mid-July to mid-August.

Blue Copper
Lycaena heteronea
Wingspan: 30–35 mm

The Blue Copper not only looks like a blue, it acts like one, too. In contrast to other coppers, the male Blue Copper does not perch and wait for females to fly close by. Instead, he patrols on the wing the way blues do. This is what biologists call convergence, the process by which distantly related lineages of organisms evolve to resemble one another in one or more aspects. For a copper to be blue in colour is interesting, but the story doesn't end there. Even the food plant of this butterfly is unusual because many of the blues, but none of the other coppers other than the Blue Copper, feed on buckwheat as caterpillars. Is there a connection? Well, probably not. There is no reason to believe that what a butterfly eats as a larva

Similar Species

Boisduval's Blue

determines whether it becomes a percher or a patroller in its courtship behaviour, or whether patrolling predisposes butterflies to feed on buckwheat as larvae. To the practiced eye, the Blue Copper is clearly not a blue—there is something about the black veins, the proportions of the wings and the overall appearance of the butterfly in flight that distinguishes it as a copper. Having said this, it is important to note that Samuel Scudder, who described this species in the first place, thought that it was a blue. As well, keep in mind that the blue and orange colours on butterfly wings are not created by blue and orange pigments. Instead, the blue colour is "structural" and results from the reflection of either blue or orange light. Thus, all that is required to see blue colour is a shift in the wavelength of the iridescence (and many coppers already have a blue-purple sheen, even on top of yellow-orange pigments).

Also Called: *Chalceria heteronea*.

ID: the only copper with blue on the upper-wing surfaces; underwings are very pale with a few small, dark spots. *Male:* bright blue on the topside. *Female:* blue at the base of the wings.

Similar Species: likely to be confused with the blues only, all of which are smaller. *Boisduval's Blue* (p. 176): is most frequently misidentified as a Blue Copper.

Caterpillar Food Plants: buckwheat (Polygonaceae: *Eriogonum* spp.).

Habitat & Flight Season: sunny hillsides, pine woodlands and sagebrush areas, usually near the food plant; flies in late June to mid-August, but later at higher elevations.

Dorcas Copper
Lycaena dorcas
Wingspan: 22–30 mm

This species and the Purplish Copper are so similar that it may, at times, be impossible to tell one from the other. In general, most butterfly experts now believe that if the caterpillar feeds on shrubby cinquefoil and the population goes through only one generation per year, we are looking at the Dorcas Copper. If the larva feeds on other plants, and the population exhibits two generations per year, it is a Purplish. However, some specialists find that there are populations in British Columbia that seem to be intermediate between the two, and for which we do not have accurate enough field observations to identify for certain which species is involved. The two

Similar Species

Purplish Copper

species do not, however, seem to interbreed.

The odd name "Dorcas" apparently refers to either a gazelle or to the roe deer of Europe, possibly because of the warm colour of the hind underwing. At a distance, this wing surface can look a tan colour in both the Dorcas and Purplish Coppers, but up close, it becomes a complex mix of violet, grey, orange and even yellow. It's too bad that the name "Dorcas" makes people laugh and call this the "Dorky Copper," but hey, that's human nature. When the name Mustard White was in use many of my friends would often call these butterflies "Mouse-turd Whites" which was worse, I suppose. Let's face it, we can't be sensitive and reverential all the time!

Also Called: *Epidemia dorcas.*

ID: a smallish copper with a tan underwing ground colour; thin orange zigzag band near the outer edge of the hind wing. *Male:* upperwing is purplish brown. *Female:* upperwing has more orange.

Similar Species: *Purplish Copper* (p. 124): very similar, but larger; more extensive orange band on the upper hind wing; more pale-coloured area on the upperwings in general.

Caterpillar Food Plants: shrubby cinquefoil (Rosaceae: *Dasiphora fruticosa* ssp. *floribunda* [syn. *Potentilla fruticosa*, *Pentaphylloides floribunda*]).

Habitat & Flight Season: wet clearings in the boreal forest and in peatlands; flies from mid-July to early August.

Purplish Copper
Lycaena helloides
Wingspan: 25–35 mm

♂

The Purplish Copper, despite its similarity to the Dorcas, is a much more familiar butterfly to most people. In fact, it is probably fair to say that this is the copper that comes most quickly to mind when naturalists in British Columbia hear the word. Part of the reason for this is that, unlike the Dorcas, the Purplish Copper is not afraid to wander far from its food plants. Of course, butterflies are probably not afraid of

Similar Species

Dorcas Copper

Bronze Copper

Purplish Copper

anything in the human sense of the word. But if it were to be discovered that butterflies had the capacity for courage, coppers would certainly rank among the bravest of the lot, since they routinely attack butterflies (and other creatures) many times their size.

In its second brood, this species can survive well into the latter part of the butterfly season. This was brought home to me in a most pleasant way when, in the process of writing this account, I took a break and walked to a nearby restaurant for lunch. On the walk back, on a lovely, warm October 1st, I came upon a fresh female Purplish Copper, flittering about between the sidewalk and a neighbour's garden.

Also Called: *Epidemia helloides*.

ID: a smallish copper with a tan underwing. *Male:* mostly brownish on the topside, with a purplish sheen, especially when young. *Female:* much more orange on the upperwing.

Similar Species: *Bronze Copper* (p. 116): male is similar but is larger, less heavily spotted with black, and has a wider orange band. *Dorcas Copper* (p. 122): slightly smaller with a smaller orange band and less extensive light areas on the wings.

Caterpillar Food Plants: sorrel, dock, smartweed and knotweed (Polygonaceae: *Rumex* spp. and *Polygonum* spp.).

Habitat & Flight Season: weedy fields and wet meadows, as well as a variety of early-successional disturbed habitats, including lawns and vacant lots; flies in two broods, the first mainly in June, the second from mid-August to mid-September.

Lilac-bordered Copper

Lycaena nivalis

Wingspan: 30–35 mm

This butterfly is relatively rare and localized in British Columbia, and it deserves the attention it gets from butterfly enthusiasts. The purplish colour on the underwing is most visible on young adults, but it is still the best identification feature. Watch for the Lilac-bordered Copper in the appropriate habitats, and especially

Similar Species

Dorcas Copper

Purplish Copper

Lilac-bordered Copper

near flowering buckwheat plants, one of its favourite nectar sources.

I feel compelled here to admit that I have never seen a Lilac-bordered Copper in the field. The reason I write this is simply to remind you that field guide authors are not all-knowing, and that people like me look forward to new experiences with new butterflies, the way you yourself might. That doesn't mean I don't have valuable things to say, but it does amaze me how many people believe that books are written only by those who have seen, done, thought through and resolved all possible aspects of their chosen passion.

Also Called: *Epidemia nivalis*.

ID: the underside of the hind wing is lilac purple on the outer third, with very small dark spots, and appears distinctly two-toned. *Male:* mostly brownish with a purplish sheen. *Female:* brownish with dark spots on the topside; no yellow or orange areas.

Similar Species: this species is most similar to its close relatives, the *Dorcas Copper* (p. 122): and *Purplish Copper* (p. 124), but is easy to distinguish by the hind wing colour.

Caterpillar Food Plants: probably the same as their close relatives, sorrel, dock, smartweed and knotweed (Polygonaceae: *Rumex* spp. and *Polygonum* spp.).

Habitat & Flight Season: relatively dry meadows, including those within forests or along stream sides; flies from late May to mid-August; most abundant in July.

Mariposa Copper
Lycaena mariposa
Wingspan: 25–30 mm

The Mariposa Copper is a great little butterfly, with a very uncopper-like pattern on its underwings that makes it, thankfully, easy to recognize in the field. You will notice that it is related to the Dorcas, Purplish and Lilac-bordered Coppers in the subgenus *Epidemia*. Some lepidopterists prefer to use this name as a full genus, and

Similar Species

Dorcas Copper

Purplish Copper

Lilac-bordered Copper

Mariposa Copper

restrict the use of the name *Lycaena* to the American Copper and its relatives. Most of us, however, prefer to use the name *Lycaena* in the broad sense, to indicate the relatedness of all of the coppers while keeping the subgenus names as such to indicate clusters of related species within the coppers as a whole. The Mariposa (whose name means "butterfly" in Spanish) is the odd member of the *Epidemia* clan. It not only has a unique underwing, it also feeds on a different family of plants as a caterpillar.

Also Called: *Epidemia mariposa*.

ID: hind underwing has a mottled grey pattern; upperwings have checkered, black and white fringes. *Male:* very dark purplish brown above; checkered fringe; slight orange hind wing band. *Female:* yellow and brown above, with dark spots; thin, wavy orange hind wing band; best recognized by the checkered fringe.

Similar Species: the underside is unique. *Dorcas Copper, Purplish Copper* and *Lilac-bordered Coppers* (pp. 122–26): upper surfaces of the wings are similar.

Caterpillar Food Plants: blueberry and bog rosemary (Ericaceae: *Vaccinium* spp. and *Andromeda* spp.).

Habitat & Flight Season: clearings in forests, including streamsides; flies from late June to early September.

Hairstreaks and Elfins
(Subfamily Theclinae)

Hedgerow Hairstreak

Hairstreaks are less frequently seen than most other butterflies, but they are interesting once you take the time to find them. Male hairstreaks perch rather than patrol to find mates, and as a group, the hairstreaks are fast fliers and take nectar readily from flowers. Hairstreaks take their group name from the thin, pale streak on the underside of most species' hind wings. Most but not all have a tiny "tail" on the hind wing as well, and it is typical of hairstreaks to perch upside-down and rub the two tails together, such that they look like antennae. Enhancing this effect, there is also a spot at the base of the tails that looks a bit like a head. Birds often peck at this "thecla spot" and on many aging hairstreaks, this area of the wing has been torn away. *Thecla* refers to a virgin and martyr in Roman mythology and was the genus name for all of the hairstreaks in North America until about a hundred years ago.

Hairstreaks comprise the largest subgroup of the gossamer-winged butterflies. Their larvae are typical

Hoary Elfin (above), Gray Hairstreak (below)

of the family, and possess honey glands (unlike coppers, which do not), but not the eversible tubules that are seen on the larvae of blues. As well, the males can be characterized by an obvious patch of androconial scales near the leading edge of the upper forewing, from which courtship pheromones are released. Hairstreaks are most abundant and diverse in the American tropics, and the diversity of hairstreaks increases steadily as one proceeds south from Canada to the equator.

Coral Hairstreak

Satyrium titus

Wingspan: 30–35 mm

This is one of the easiest hairstreaks to identify. In 1904, lepidopterists John H. Comstock and his wife Anna could not resist quoting a 19th-century lepidopterist, John Abbot ("the gentle Abbot"), who described this species as "'a little brown butterfly,' which, though not distinctive, is deeply satisfactory when one knows the species" (*How to Know the Butterflies*, p. 235). I can't resist requoting that here; it captures so many people's feelings about the Coral Hairstreak. Finding one, typically perched on chokecherry leaves, is always a happy occasion. They sit with their wings closed over their backs, and their "coral" spots in full view. (The spots are more pinkish and coral-coloured in eastern North America.)

To those of us who have been at this butterfly thing for a while, it is a bit of a shame that the Coral Hairstreak is no longer placed in the genus *Harkenclenus*. Our fondness for the name comes not just from a resistance to change (the reasons for calling it a *Satyrium* are good ones), but because *Harkenclenus* is a clever abbreviation. Harold Kenneth Clench was a great lepidopterist who, although I never met him, was also a good friend to many of my older colleagues. The name *Satyrium* has its own pizzazz, mind you. It refers to the voluptuous, goat-like satyrs, woodland gods who were associated with Bacchus (the god of wine and intoxication) in Greek mythology. To many of us, then, finding a Coral Hairstreak is much more than a chance encounter with a brown butterfly—it brings a whole host of images to mind!

Also Called: *Harkenclenus titus*.

ID: the wings are mostly brown-grey, on the topside and underside; string of orange spots on the outside rim of the hind wing underside; sexes are similar.

Similar Species: this species is distinctive.

Caterpillar Food Plants: known to feed on cherry trees in B.C. (Rosaceae: *Prunus* spp.); may also feed on saskatoon (Rosaceae: *Amelanchier alnifolia*) or oak (Fagaceae: *Quercus* spp.).

Habitat & Flight Season: shrubby places along river valleys and streams, as well as in meadows and along roadsides; flies from late June to early August, with the possibility of a few second-brood butterflies emerging in the fall.

Behr's Hairstreak

Satyrium behrii

Wingspan: about 28–30 mm

The Behr's Hairstreak is not a particularly showy butterfly, but it is interesting nonetheless. In British Columbia, its future is not entirely secure, and the antelope brush slopes it lives on are threatened by both housing developments and agriculture. Also, changes in range management may reduce the amount of antelope brush available to the caterpillars. The fact remains that just about any species with specialized habitat requirements, and a range restricted to the southern Okanagan in B.C., is probably in some danger of population decline.

This butterfly was named for Herman Behr, a lepidopterist at the California Academy of Sciences in the mid-1800s, the time many of us think of as Darwin's day. Behr is the namesake of a number of North American butterflies, but usually at the subspecies level. He also named many of our butterflies (technically that is—he did not coin English names for the butterflies he discovered), including *Euphilotes battoides*, the Square-spotted Blue.

Similar Species

Mariposa Copper

ID: no tails; wings are orange on the topside and broadly bordered with black; hind wing underside is distinctive, but looks much like that of a blue; sexes are similar.

Similar Species: *Mariposa Copper* (p. 128): female looks somewhat similar on the topside, but has a checkered fringe and a blotchier pattern of orange on the upperwing surfaces.

Caterpillar Food Plants: antelope brush (Rosaceae: *Purshia tridentata*) in B.C., and mountain mahogany (Rosaceae: *Cerocarpus montanus*) farther south in the U.S.

Habitat & Flight Season: dry slopes with antelope brush in the southern Okanagan; flies in June and into July.

Sooty Hairstreak
Satyrium fuliginosa

Wingspan: 25–30 mm

The Sooty Hairstreak is like the Blue Copper in that it is said to be convergent with true blues (or true-blue blues, I suppose you could say). But unlike the Blue Copper, the Sooty Hairstreak shows no blue colour and instead looks like a brown female blue. It is on the underwing that the resemblance is most striking, and it may well be that this familiar underwing pattern is much like the long-gone common ancestor of all the gossamer-winged butterflies. If so, a few evolutionary possibilities come to mind. On the one hand, it is possible that the blue-style underwing pattern was retained by the Blue Copper, the Sooty Hairstreak and the true blues, and evolved into other patterns in the other gossamer-winged butterfly lineages. On the other hand, it may be that these three sorts of butterflies came to resemble each other independently, in which case the blue-style pattern, even if it was the "primitive" pattern, was reacquired by the Blue Copper and the Sooty Hairstreak (a rare example of evolution reversing itself). This shows, if nothing else, how interesting the study of butterfly relationships can be, above and beyond the mere naming of species. It is worth pointing out, as well, that lupines are an unusual food plant for hairstreaks, but are commonly used by some blues. But before you get the impression that we know this butterfly in detail, keep in mind that the Sooty Hairstreaks' egg and pupal stages are still completely unknown to science.

Also Called: Sooty Gossamer Wing; *Satyrium fuliginosum*.

ID: the wings are grey-brown on the topside, and pale, but spotted like a blue's, underneath; small for a hairstreak; unusually rounded wing shape; sexes are similar.

Similar Species: most likely to be confused with a female blue, especially species with larger dark spots on the front wings underside compared to the hind wings underside, such as the *Boisduval's Blue* (p. 176); and *Arctic Blue* (p. 182).

Caterpillar Food Plants: the larvae feed on lupines (Fabaceae: *Lupinus* spp.).

Habitat & Flight Season: found in open areas, including sagebrush, where lupines grow; flies mostly in June, with some remaining in May and July.

California Hairstreak

Satyrium californica

Wingspan: about 30–35 mm

This is another hairstreak of the southern Interior, and not surprisingly it is of concern to butterfly conservationists. The plight of this species in the province is confounded by the difficulty lepidopterists have historically had in separating it from the Sylvan Hairstreak. For example, there are records of California Hairstreak larvae feeding on willow leaves in B.C., but some lepidopterists believe these probably refer to the Sylvan Hairstreak instead. As well, some believe that these two species hybridize from time to time, while others see them as a reproductively separate species. In instances like this, it is extremely difficult to assess how a species is doing, from a preservationist point of view. For one thing, it is difficult to monitor a species that you may or may not be able to recognize. For another, it is difficult to know whether the "species" is worth protecting because it may or may not represent a real species. For those who find the picky details of butterfly taxonomy frustrating, this is as good an example as any that details matter, no matter how annoying they may seem at first. Personally, I have always found taxonomy fascinating, and I have no great sympathy for those who find it an unfathomable aspect of the study of life.

Similar Species

Sylvan Hairstreak

Also Called: *Satyrium californicum*.

ID: grey topside with orange near the tails and the outer hind corner of the forewing. Hind underwings with blue, orange-capped thecla spot and two rows of black dots, the outer row capped with orange.

Similar Species: *Sylvan Hairstreak* (p. 136): similar, but no orange cap on the blue thecla spot; paler grey on the underside; less orange on the outer row of spots.

Caterpillar Food Plants: uncertain in B.C.; some records show it eats saskatoon (Rosaceae: *Amelanchier alnifolia*), willow (Salicaceae: *Salix* spp.), oak (Fagaceae: *Quercus* spp.), chokecherry (Rosaceae: *Prunus virginiana*), mountain mahogany (Rosaceae: *Cerocarpus montanus*) and antelope brush (Rosaceae: *Purshia tridentata*).

Habitat & Flight Season: open dry areas, including hillsides; flies in June, July and early August.

Sylvan Hairstreak
Satyrium sylvinus
Wingspan: about 25–35 mm

Similar Species

California Hairstreak

The word "sylvan" means "of the forest" and indeed this is more of a forest butterfly than its close relative, the California Hairstreak. But, don't let this give you the wrong impression; forests, *per se*, are not to its liking. The Sylvan Hairstreak generally flies in reasonably open habitats with relatively high soil moisture, and is thus naturally associated with willows. This type of habitat serves this butterfly well in British Columbia, and may explain why it is the most abundant and widely distributed member of its genus in the province. The others (the Coral, Behr's, Sooty, California, Striped and Hedgerow Hairstreaks) seem to prefer drier places and, perhaps coincidentally, are less widely distributed. As well, Sylvan Hairstreaks often visit milkweed flowers for nectar, and milkweed's native habitat does not usually include wet places. "Open," as used to describe butterfly habitats, means simply this: from the perspective of a standing adult human being, there are few if any tall plants in the way when one looks around the immediate area in search of butterflies. From a butterfly's point of view, I'm sure the top of a tree is equally "open," and the most "open" habitats of all are sand dunes and mud flats.

Also Called: *Satyrium sylvinum.*

ID: grey on the topside, with orange patches near the tails; hind underwings have an orange-capped, blue thecla spot with an inner row of black dots with white halos and an outer row of black-capped, orange dots; sexes are similar.

Similar Species: *California Hairstreak* (p. 135): very similar, but with more orange overall, especially over the thecla spot.

Caterpillar Food Plants: willows (Salicaceae: *Salix* spp.).

Habitat & Flight Season: moist meadows, clearings and streamsides; flies mainly in July, with records ranging from late June to early August.

Striped Hairstreak
Satyrium liparops
Wingspan: about 25–30 mm

The Striped Hairstreak is one of my favourite butterflies. In British Columbia, it is found only in the Peace River area, as a population, isolated from its relatives in the aspen parklands farther south. In the opinion of some lepidopterists, the Peace River populations belong to the *fletcheri* race of the species, and the populations farther south belong to the *liparops* race. Others prefer to call both the aspen parkland populations and those in the Peace River area *fletcheri*. This division illustrates an interesting pattern, which is also seen in a number of other butterflies. The Peace River area is home to parklands and grasslands that are isolated from similar habitats to the south. The isolated populations of butterflies living in the Peace River area have probably been isolated from their relatives for at least 5000 years, which may well be long enough for them to have evolved distinctive characteristics. We know that the Peace River populations of the Old World Swallowtail (the *pikei* race) are distinctive, but no one has yet studied the Striped Hairstreaks to the same degree.

Also Called: *Strymon liparops*.

ID: grey-brown on the topside; orange patches on the forewings and near the tails; thin, short, roughly parallel lines across the grey-brown background on the underside; small, but obvious blue and orange thecla spot; sexes are similar.

Similar Species: the other *Satyrium* hairstreaks show their relatedness to the Striped Hairstreak, but none shares the underwing pattern characteristic of this species.

Caterpillar Food Plants: possibly saskatoon (Rosaceae: *Amelanchier alnifolia*) and cherry (Rosaceae: *Prunus*) shrubs, as well as oak willow, poplar and blueberry, all of which have been recorded as host plants in eastern Canada.

Habitat & Flight Season: shrubs, especially saskatoon and chokecherry; flies from late June to late July.

Hedgerow Hairstreak
Satyrium saepium
Wingspan: about 28–32 mm

This is the last of the *Satyrium* hairstreaks on our list, and a nice one at that. It is reasonably easy to identify, and although it does not exhibit any spectacular characteristics, it is still uncommon and localized enough that it is always a pleasure to encounter. The name "Hedgerow" comes from the same source as the word "*saepium*," which also refers to hedges, and it is thought that it may be a reference to the ceanothus food plant.

Hedgerow Hairstreaks can be quite pretty when they first emerge, and the upperside of the wings is then quite iridescent copper in colour. This fades, however, to the point where the butterfly looks dull and brown. The British Columbia populations (along with those in adjacent Washington to the south) have been given their own subspecies name, *Satyrium saepium okanaganum*, named in 1944 and with a type locality of Peachland. They differ in relatively subtle ways from Hedgerow Hairstreaks from farther south in the western United States.

Similar Species

Juniper Hairstreak

ID: reddish brown on the topside, with a faint, pale bar at the outer end of the cell; dark grey-brown on the underside, with a single "hair-streak" line; black and blue patch near the base of the short tail; sexes are similar.

Similar Species: *Juniper Hairstreak* (p. 144): female is similar, but only on the topside.

Caterpillar Food Plants: ceanothus (Rhamnaceae: *Ceanothus* spp.).

Habitat & Flight Season: found near host plant, ceanothus, in pine forests and other relatively dry, relatively open areas; flies in June, July and August.

Bramble Hairstreak
Callophrys affinis
Wingspan: about 22–30 mm

What a shame that such a delicately beautiful butterfly should lie at the centre of such a taxonomic mess! Among those attempting to match names to the "green hairstreaks" there has been a long history of confusion and conflict, the result of which has been the recognition of between two and seven species. To make things even more difficult to group together, many of these taxa (a word referring to any named group in biology) intergrade, and the butterflies themselves probably hybridize. To top it all off, some lepidopterists use the genus name *Callophrys* to refer to the green hairstreaks alone (the way it is used here), while others use it in a much broader sense to include the elfins, and the Thicket, Johnson's, Cedar and Juniper Hairstreaks. This is not, however, sufficient grounds for assuming that we can arbitrarily name these butterflies. The *affinis* taxon, be it a species or subspecies, is "real" at some level of abstraction, and one of the things that is real about it is that it is not very common in British Columbia. It occurs in localized, smallish populations, and in government circles is considered a "species of concern" (an official government designation) with respect to conservation.

Similar Species

Sheridan's Hairstreak

Also Called: Immaculate Green Hairstreak, Immaculate Bramble Hairstreak or Western Green Hairstreak; *Callophrys dumetorum affinis* or *Callophrys perplexa affinis*.

ID: grey on the topside and green on the underside; a few white flecks are more or less arranged as a hairstreak line, although some have no white at all. *Male:* slightly greyer on the topside. *Female:* slightly browner on the topside; sexes are similar.

Similar Species: Sheridan's Hairstreak (p. 140): similar, but has a more prominent white hairstreak, and flies earlier in the season.

Caterpillar Food Plants: buckwheat (Polygonaceae: *Eriogonum* spp.).

Habitat & Flight Season: relatively dry, relatively open areas, usually with buckwheat plants; flies from May to mid-June.

Sheridan's Hairstreak
Callophrys sheridanii
Wingspan: about 22–26 mm

This is the second green hairstreak we will discuss, but in terms of its emergence during the season, it is the first. Despite what I wrote about the taxonomic confusion surrounding the green hairstreaks under the preceeding species, it is reasonably clear that the two British Columbian butterflies, the Bramble and the Sheridan's, are separate at the species level. That is to say, they do not interbreed, or at least not within the borders of our province. Of course, you might want to keep this in mind if and when you travel farther afield in the Pacific Northwest. I would hate to think that I am sending Canadian butterfly enthusiasts out into the world holding "provincial" views of the complexity of life.

The name Sheridan, by the way, comes from our lepidopterist colleagues in the United States. In 1877, entomologist/taxonomist W.H. Edwards named this butterfly in honour of Lieutenant General P.H. Sheridan, an American Civil War hero. Now is that an odd thing to do to a little green butterfly, or what? I personally can't help thinking of the British television comedy "Keeping Up Appearances," and the mysterious son, Sheridan, who is continually calling his doting mother to beg for more money. This is the price of language—it will call to mind whatever it calls to mind!

Similar Species

Bramble Hairstreak

Also Called: White-lined Sheridan's Hairstreak.

ID: grey-brown on the topside, and green on the underside, with a white hairstreak line across the underwings; sexes are similar.

Similar Species: *Bramble Hairstreak* (p. 139): has a less prominent (or non-existant) white line, although the line is variable in both species.

Caterpillar Food Plants: buckwheat (Polygonaceae: *Eriogonum* spp.).

Habitat & Flight Season: open areas with buckwheat plants and often sagebrush; range also extends to higher elevations; flies in April and May at low elevations, and June and July up higher.

Thicket Hairstreak

Callophrys spinetorum

Wingspan: about 25–30 mm

Compared to the hairstreaks we have discussed previously, this one has truly odd habits. The Thicket Hairstreak spends most of its time high in trees near its mistletoe food plants, and comes down to ground level only for the occasional visit to nectar flowers.

In the scientific names that have been given to this butterfly, we see the distinction between splitting and lumping. *Loranthomitoura*, "splits" the Thicket Hairstreak and its close relatives away from the other *Mitoura* hairstreaks (for example, the Juniper Hairstreak) with which it is sometimes placed (or "lumped"). Even if we used the name *Mitoura*, we would be splitting this whole group of hairstreaks away from the larger group that was traditionally called *Callophrys* (and before that, *Thecla*). Here, I choose to follow modern usage and treat all of these names as synonyms of *Callophrys*. To understand this, all you have to remember is that each name is still useful, but in some books it will appear as a genus, while in others it is a subgenus, or a genus group. And as for the misleading name "Thicket Hairstreak"? It comes from the resemblance of this species to its thicket-loving relatives in Europe, and the Latin word *spinetum*, meaning thickets.

Similar Species

Johnson's Hairstreak

Also Called: *Mitoura spinetorum* or *Loranthomitoura spinetorum*.

ID: blue on the topside, smeared into dark grey near the edges of the wings; brown with a white hairstreak line, that forms a "W" over the orange and blue thecla spot on the underside; occasionally there is a pale bar at the end of the cell on the forewing underside. *Male:* a bit more blue on topside. *Female:* a bit more grey on topside.

Similar Species: *Johnson's Hairstreak* (p. 142): almost identical on the underside, but brown on the topside; has no pale cell bar.

Caterpillar Food Plants: dwarf mistletoes (Visaceae: *Arceuthobium* spp.) growing on coniferous trees.

Habitat & Flight Season: thickets? Of course not. Often in open pine forests; flies from late April to mid-July.

Johnson's Hairstreak

Callophrys johnsoni

Wingspan: about 30–35 mm

This is a rare butterfly in British Columbia, and according to *The Butterflies of Canada*, there are only four colonies remaining in the province. These are located on southeastern Vancouver Island and in the lower Fraser Valley, and although there are a lot of western hemlocks in these areas, the Johnson's Hairstreak is still quite uncommon. Lepidopterists Crispin Guppy and Jon Shepard speculate that mistletoe eradication may be partly to blame for the rarity of this butterfly, and this may well be true. It may also be the case that spraying the moth-killing pesticide Btk for gypsy moths may have had a heavy, negative impact on Johnson's Hairstreak. On the other hand, it is apparently sometimes common in California, and we can always hope that it stages a comeback some time in the future, or that new populations will be discovered. It is, after all, a tiny, brown butterfly that spends most of its time at the tops of trees. This species was named for a Professor O.B. Johnson, of Washington State University, by Henry Skinner in 1904.

Similar Species

Thicket Hairstreak

Also Called: *Mitoura johnsoni* or *Loranthomitoura johnsoni*.

ID: brown overall with a white hairstreak line that forms a "W" over the orange and blue thecla spot on underside. *Male:* deeper brown on the topside. *Female:* more reddish brown on the topside.

Similar Species: *Thicket Hairstreak* (p. 141): has a blue topside. *Cedar Hairstreak* (p. 143) and *Juniper Hairstreak* (p. 144): both have a clearly delineated dark border on the topside, and a bit of a greyish cast on the hind wing underside.

Caterpillar Food Plants: dwarf mistletoes (Visaceae: *Arceuthobium* spp.) growing on western hemlock.

Habitat & Flight Season: western hemlock forests with mistletoe plants; flies in late May to early July.

Cedar Hairstreak
Callophrys nelsoni
Wingspan: about 25–30 mm

Sheesh—look at all those names! But, sorting it all out is relatively simple. First, there is the question of what genus name to use, which is determined by whether one wants to emphasize similarities among a large group of species (as a "lumper") or differences among numerous smaller groups of species (as a "splitter"). This is discussed previously, under the Thicket Hairstreak (p. 141).

Then there is the question of which specific epithet to place with the genus name, and this question has two parts. First, there is the issue of priority, meaning that the name that was first proposed for the species (and hasn't been used for any other species) is the one we should use. Then there is the issue of whether a particular name should apply to a full species or a subspecies (a geographic race). In the case of the Juniper Hairstreak, I am following those who believe that it belongs in the large genus *Callophrys* and the species *C. nelsoni*.

Similar Species

Juniper Hairstreak

Also Called: Rosner's Hairstreak, Rosner's Juniper Hairstreak, Nelson's Juniper Hairstreak or simply as part of the Juniper Hairstreak; *Mitoura rosneri, Callophrys rosneri, Mitoura nelsoni*, part of *Callophrys siva, Mitoura siva, Callophrys gryneus, Mitoura gryneus* or *Callophrys grynea*.

ID: orange-brown on the topside with black outer wing borders, and an iridescent purple sheen when young; brown on the underside; more orange on the forewing and grey-purple on the hind wing; obvious hairstreak line on the underside; male has more orange on the forewing than the female.

Similar Species: *Juniper Hairstreak* (p. 144): shows more contrast between the base and outer portions of the hind wing undersides; lives on Rocky Mountain juniper, not cedars.

Caterpillar Food Plants: the caterpillars eat western red-cedar (Cupressaceae: *Thuja plicata*), but they may also eat junipers (Cupressaceae: *Juniperus* spp.).

Habitat & Flight Season: usually in the vicinity of western red-cedar, but colonies are localized and relatively scarce; flies from late April to early July, with a longer flight season near the coast.

Juniper Hairstreak
Callophrys gryneus
Wingspan: about 25–30 mm

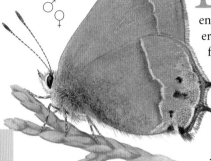

Despite its being almost identical to the Cedar Hairstreak, the Juniper appears to show consistent differences, both in its wing pattern and its preference for junipers over cedars as a larvae food plant. At least that is the opinion of those who know the butterfly best.

Apparently, our B.C. subspecies of this butterfly (*C. gryneus siva*) was named for the Sivas area of Turkey. Why? I have no idea. The person who described it was entomologist and taxonomist W.H. Edwards, in 1874, who should have known that it lived nowhere near Turkey. But strange things happened in the 1800s. For example, one of our most common tiger beetles, *Cicindela tranquebarica*, was named by another entomologist for the town of Tranquebar in India. For many museum entomologists, it was a time when specimens were coming in from all over the world, awaiting formal descriptions and new scientific names. Anyone who expects scientific names to make sense or possess profound meaning is bound to be disappointed.

Similar Species

Cedar Hairstreak

Also Called: Siva Hairstreak, Barry's Juniper Hairstreak or Siva Juniper Hairstreak; *Mitoura barryi, Callophrys siva, Callophrys grynea, Callophrys grynea siva* and so on...

ID: orange-brown on the topside, with black outer wing borders, and an iridescent purple sheen when young; brown on underside, with more orange on the forewing and grey-pink on the hind wing; obvious hairstreak line on underside; male has more orange on the forewing underside than the female.

Similar Species: *Cedar Hairstreak* (p. 143): more violet brown (not greenish grey) when fresh; has a less contrasting pattern (there is more difference between adjacent light and dark areas), less pinkish hind wing underside.

Caterpillar Food Plants: Rocky Mountain juniper (Cupressaceae: *Juniperus scopulorum*).

Habitat & Flight Season: adults are found on and around the juniper host plants; flies in mid-April through May.

Brown Elfin
Callophrys augustinus
Wingspan: about 20–28 mm

The Brown Elfin is a widespread, common butterfly throughout most of Canada, but in British Columbia it is found only in the northeast corner of the province, where few naturalists have a chance to see it. Not that I'm complaining—we have a rather diverse elfin fauna in B.C., with six species in all. Early in the butterfly season, after the anglewings, tortoiseshells

Similar Species

Moss's Elfin   Western Elfin   Hoary Elfin

and Mourning Cloaks have emerged from hibernation, and the Spring Azure blues are on the wing, elfins give butterfly enthusiasts something a bit more challenging and elusive to search for. When you first encounter elfins, it will be immediately obvious that they are gossamer-winged butterflies, by their dark eyes, white eye-rings and small size. It may, however, not be apparent at first that they are members of the hairstreak subfamily; but if you watch them for a while, you will notice that they are less flighty than the blues and less flamboyant than the coppers. I like their names, too. "Elfin" evokes an image of the coniferous forests in which they live, and *augustinus* is a reference to Augustus, an Inuit who traveled with the expeditions of Sir John Franklin through the Canadian Arctic. Sir John Richardson, the ship's naturalist, wrote of Augustus' contributions to the expedition, but it was John O. Westwood, in 1852, who named the Brown Elfin in Augustus's honour.

I have had the pleasure of getting to know this species well in Alberta, where it is the most common elfin in most places. The Rocky Mountains are a profound barrier to butterfly distribution, and it is wonderful that British Columbians can often see their dream butterflies in Alberta, and Albertans can do the same in B.C.

Unlike some of the blues, I find it impossible to tell elfins apart on the wing. Perhaps this is because they fly in typical hairstreak fashion, fast and close to the ground, rather than lazily fluttering through the air like a blue.

Also Called: *Incisalia augustus* or *Incisalia augustinus*.

ID: a small, brown, tailless hairstreak, typical of the elfin group, which is sometimes not recognizable as a hairstreak at first glance; topside of wings is uniform brown; underside is two-tone brown, lighter on the outer half of all four wings; sexes are similar.

Similar Species: *Western Elfin* (p. 147): similar, but more reddish underneath. *Moss's Elfin* (p. 148) and *Hoary Elfin* (p. 149): the hind wing undersides are a frosty grey (whereas the two pine elfins are much more angular).

Caterpillar Food Plants: hosts may include bearberry (Ericaceae: *Arctostaphylos uva-ursi*), blueberries (Ericaceae: *Vaccinium* spp.), labrador tea (Ericaceae: *Ledum groenlandicum*) and leatherleaf (Ericaceae: *Chamaedaphne calyculata*).

Habitat & Flight Season: near the ground in relatively sunny places with coniferous trees; flies from late April to early June.

Western Elfin

Callophrys iroides

Wingspan: about 28–30 mm

Is this just a coastal subspecies of the Brown Elfin (along with some other populations in Arizona and New Mexico), as many lepidopterists believe, or does it deserve to be recognized as its own species? I have opted for the latter arrangement, but only in order to draw attention to the possibility, not because I firmly believe one way or the other. In favour of recognizing two separate species, they differ geographically and are recognizable by wing pattern. In favour of considering the Western Elfin to be a subspecies of the Brown, it is reasonable to expect wing pattern differences within a single widespread species, but geographic separation alone does not constitute sufficient grounds for recognizing them as exclusive. One of the most interesting features of these two butterflies is that they enter the pupal stage in June and remain there, patiently, until the following spring. This is not a record, however, and many butterflies will sit out a dry year, or even a few years, before emerging. Still, it is interesting to see a butterfly that is so tightly adapted to the springtime. In B.C., this seems odd because summer and fall also support the butterfly season, but farther south it is commonly the case that spring is the only season with enough moisture for both butterflies and caterpillars to survive.

Similar Species

Brown Elfin

Also Called: Brown Elfin, in part; *Incisalia iroides*.

ID: topside of wings are uniform brown; two-tone reddish brown on the underside; sexes are similar.

Similar Species: *Brown Elfin* (p. 145): more dark chocolate brown on the underside. Other elfins have a more angular pattern on the underside, or have white frosting on the hind wing underside.

Caterpillar Food Plants: bearberry (Ericaceae: *Arctostaphylos uva-ursi*), salal (Ericaceae: *Gaultheria shallon*), arbutus (Ericaceae: *Arbutus menziesii*), ocean spray (Rosaceae: *Holodiscus discolor*) ceanothus (Rhamnaceae: *Ceanothus sanguineus*) and apple (Rosaceae: *Malus* spp.).

Habitat & Flight Season: open coniferous forests, near the ground; flies from late April to early June.

Moss's Elfin
Callophrys mossii
Wingspan: about 25–30 mm

The lovely little Moss's Elfin was named in the 1880s for Mr. David Moss of Esquimault, British Columbia. It is very much a western species, and there is only one record of Moss's Elfin to the east in Alberta, over the Continental Divide. In the interior of British Columbia, the species seems to be in good ecological shape, but apparently the Vancouver Island populations live in habitats that are decreasing in quality. B.C. lepidopterists Crispin Guppy and Jon Shepard feel that the slopes they require are threatened not only by development, but also deer grazing and the erosion caused by climbers and hikers. Thus, the Moss's Elfin adds its name to the long list of plant and animal species that are surviving less and less well on the island, despite what a wonderful place we all know it to be.

The authors that name this butterfly species *Callophrys fotis* are in the minority. Most feel that the latter is restricted to the Great Basin of the United States, and feeds on an entirely different food plant. Of course, this is not a foolproof measure of species status, any more than wing pattern or geographic separation might be, so it is likely that differences of opinion will persist on this subject.

Similar Species

Hoary Elfin

Also Called: *Incisalia mossii* or *Callophrys fotis mossii*.

ID: generally brown, with a two-toned hind wing on the underside, with brown at the base and blotchy grey on the outer half; white wing fringe; sexes are similar.

Similar Species: *Hoary Elfin* (p. 149): distinguished by a more prominent white line dividing the inner and outer halves of the hind wing underside.

Caterpillar Food Plants: at least two species of stonecrop (Crassulaceae: *Sedum spathulifolium* and *S. lanceolatum*).

Habitat & Flight Season: found on slopes, usually rocky; flies from March through June.

Hoary Elfin
Callophrys polios
Wingspan: 22–30 mm

Found throughout British Columbia to the east of the Coast Ranges, this is probably the elfin most familiar to naturalists in the province. It is also relatively easy to identify, making it an easy species to recognize and remember. But it is also a typical elfin in almost every way, emerging in the spring, staying near the ground and frequenting sunny places in or near coniferous forests. Elfins, as well, almost never spread their wings to bask, so the underwing patterns become all the more important to butterfly enthusiasts. To my eye, the frosted look of the Hoary Elfin's hind wing adds to its camouflage, and these butterflies can be downright tricky to spot if you don't see them moving. In eastern Canada and the northeastern United States, there are other "frosted" elfins, such as the Henry's Elfin and the Frosted Elfin. These latter two butterflies are "tailed," making their relationship to the hairstreaks that much easier to notice. Among our six species of British Columbia elfins, it is handy to remember that two are brown, two are frosted and two have a more angular pattern on the underwing.

Similar Species

Moss's Elfin

Also Called: *Incisalia polia*, *Incisalia polios* or *Callophyrus polia*.

ID: a small elfin; brown on the topside and two-toned on the underside; hind wing underside has a strongly frosted appearance on the outside of the wing; sexes are similar.

Similar Species: *Moss's Elfin* (p. 148): has a less prominent white line separating the inner and outer halves of the hind wing on the underside.

Caterpillar Food Plants: bearberry (Ericaceae: *Arctostaphylos uva-ursi*).

Habitat & Flight Season: near the ground in or near open, coniferous forests; flies in the spring from mid-April to early June.

Eastern Pine Elfin
Callophrys niphon
Wingspan: about 25–30 mm

This is a rare butterfly in British Columbia, found only in the northeast along part of the Liard Highway, in jack pine forests. Jack pine is an eastern tree that is replaced in western Canada by the lodgepole pine; in turn, the Eastern Pine Elfin is replaced by the Western Pine Elfin, which, not coincidentally, feeds on lodgepole pine needles as a caterpillar. And do the two butterfly species interbreed where the two ranges of tree species meet? This question has yet to be answered to the full satisfaction of lepidopterists, although it seems likely that they do intergrade, at least in a narrow geographic band along the east slopes of the Rockies. But, there is no solid evidence for considering them a single species. The definition of a "species" is sufficiently flexible in itself that it can accommodate some of the messiness in nature. In the end, however, it all seems to relate as much to human psychology and the need for order as it does to butterfly evolution and the species as the basic unit of nature.

Similar Species

Western Pine Elfin

Also Called: *Incisalia niphon*.

ID: brown on the topside with some orange; underside wing pattern has a zigzagged line through a brown, white and black background; two faint dark bars in the forewing cell; male is less orange on the topside than the female.

Similar Species: *Western Pine Elfin* (p. 151): has only one dark bar in the forewing cell.

Caterpillar Food Plants: jack pine (Pinaceae: *Pinus banksiana*).

Habitat & Flight Season: jack pine forests; flies in mid-June in B.C.

Western Pine Elfin

Callophrys eryphon

Wingspan: about 28–32 mm

This, our final elfin, is as widespread and familiar as the Hoary Elfin in most places. It is truly a butterfly of the West though it does occur on the eastern side of the Continental Divide. The biological and nomenclatorial distinctions between this and the Eastern Pine Elfin are discussed, under the preceeding species. Not much can be added here, except to say that the three pine trees used by the Western Pine Elfin are all considered soft pines, while the jack pine is considered a hard pine. This distinction itself has little biological significance, but it may help remember which elfin is which, if you are the sort of person who knows your wood.

Finally, it deserves mention somewhere in our discussion of elfins that this little group of butterflies is in some ways a paradox. On the one hand, they are hairstreaks, and hairstreaks are clearly a group that is tightly associated with the tropics. On the other hand, elfins are butterflies of cool, northern forests, and butterfly enthusiasts in the southern United States are always interested to hear tales of these "cold weather" creatures, which they associate with such things as moose and caribou!

Similar Species

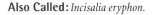

Eastern Pine Elfin

Also Called: *Incisalia eryphon.*

ID: a relatively large elfin, brown and orange on the topside; brown with an angular wing pattern on the underside; has one dark bar in the forewing cell on underside; male is less orange than the female on the topside.

Similar Species: *Eastern Pine Elfin* (p. 150): has two dark bars in the forewing cell on underside; a less regular zigzag line on the hind wing underside.

Caterpillar Food Plants: lodgepole, monticola and ponderosa pine needles (Pinaceae: *Pinus contorta, P. monticola* and *P. ponderosa*, respectively).

Habitat & Flight Season: pine forests in open, relatively sunny spots; flies from early May to early June.

Grey Hairstreak
Strymon melinus

Wingspan: about 25–30 mm

This is probably the hairstreak you should know best, and first, despite it being the last in this section. It can be encountered all across southern Canada, almost all of the United States, Mexico, Central America and parts of northern South America. Fortunately, it is also quite easy to recognize. It flies throughout the butterfly season in B.C. Many times, while chasing butterflies far from home, surrounded by unfamiliar hairstreaks, I have been delighted to find the familiar Grey. What more could we ask from such a tiny thing? In British Columbia, the Grey Hairstreak can be a pest of garden beans, but I can't imagine anyone preferring a mere bean over a few of these enchanting butterflies. As further proof that the word "pest" is a misnomer for this species, it seems to undergo very large fluctuations in population size, remaining rare for years and then appearing in abundance for a time. Recalling the words of lepidopterists John H. and Anna Comstock, who commented that the Grey Hairstreak is "a frisky little creature living up to its orange spots in action rather than to its decorous body colour" (*How to Know the Butterflies*, p. 224), I think this hairstreak should be appreciated rather than resented. Presumably, the Comstocks made different psychological associations with orange spots than we do today, but they make a good point—once you notice the Grey Hairstreak, it is a truly adorable little creature.

Also Called: Gray Hairstreak.

ID: a tailed hairstreak with grey upperwings; underwings have a black and white hairstreak line; orange thecla spot; black dashes near the wing margins; sexes are similar.

Similar Species: this is the only tailed hairstreak in B.C. with grey wings on the topside and a distinctive underside.

Caterpillar Food Plants: the caterpillars have been recorded from many plants in many families, but often found on members of the bean family, Fabaceae.

Habitat & Flight Season: a variety of habitats, including stream and river edges, forest understorey, open hilltops and gardens; flies in two broods; the first from late April to early July, the second from mid-July to early August. In the southern Interior, there can be a small third generation each year.

Blues
(Subfamily Polyommatinae)

Cranberry Blue on saskatoon berries

The blues are by far the most familiar of the gossamer-winged butterflies, primarily because of the bright blue colour of most males and their habit of patrolling on the wing in search of females. From the first Spring Azures of the season, right through to the second broods of some other species in the fall, more than a dozen species of blues make for great butterfly watching all through the season in British Columbia. Also, it is common to see male blues at "mud puddle clubs" alongside sulphurs and swallowtails.

Blues are a diverse group in temperate regions, and not in the tropics, making their evolutionary history more like that of the coppers than of the hairstreaks. As well, they frequently live in small, relatively isolated, relatively distinctive populations. This has put some types of blues at the centre of the butterfly conservation movement, ever since the Xerces Blue (*Glaucopsyche xerces*) went extinct in California. Development was to blame, and the fate of the Xerces Blue may well await some other North American species, although the blues found in our province seem safe for the moment.

The larvae of blues possess both honey glands and eversible tubules, and they eat the flowers, seeds and seedpods of their host plants, not the leaves. Adult blues are classified largely on the basis of wing veins and genitalic characters, and you will find that the classification of blues has been profoundly unstable, even in recent years. I'll do my best to make these stories interesting.

Eastern Tailed Blue

Cupido comyntas

Wingspan: about 28–30 mm

This is an interesting species in British Columbia in the sense that it may well be in danger of local

Similar Species

Western Tailed Blue

extinction (there are only three or four known populations), but may also have been recently introduced. The Eastern and Western Tailed Blues are each other's closest relatives, but it is the Western that is widespread and common in the western half of North America, appropriately enough. There is, however, an isolated part of the Eastern Tailed Blue's range that extends over part of the Pacific Northwest, and many lepidopterists

Eastern Tailed Blue

feel that this species was introduced by people to this area. Not all agree, mind you, and you will find that some feel that the tight association of the Eastern Tailed Blue with natural riparian habitats in British Columbia is evidence that it is a naturally occurring species here. (This seems like a weak argument to me, based on the concept that natural butterflies like natural places.) Others say that this butterfly only looks like the real Eastern Tailed Blue, but it is actually a separate species, that, as of this moment, has no name. Imagine, a butterfly with no name...how nice. And so many of them have too many, don't you think?

Also Called: *Everes comyntas*.

ID: a tailed blue with black wing margins on the topside, and a white fringe; there can be two or three black spots, with orange caps near the tail base on the topside; the underside is greyish. *Male:* mostly purplish blue above. *Female:* mostly brown above with some blue.

Similar Species: *Western Tailed Blue* (p. 156): the only other tailed blue that has lighter underwings, less distinct markings on the underside and a different pattern of spots near the tail base.

Caterpillar Food Plants: white clover (*Trifolium repens*), red clover (*T. pratense*) and cow vetch (*Vicia cracca*), all Fabaceae.

Habitat & Flight Season: found mostly in sunny places near rivers and streams; flies from mid-June to mid-July.

Western Tailed Blue

Cupido amyntula

Wingspan: about 28–30 mm

Despite the taxonomic confusion created by the presence of a handful of Eastern Tailed Blues in the province, the Western Tailed Blue is a well-known, not terribly variable species.

Similar Species

Eastern Tailed Blue

Butterfly enthusiasts come to know it easily, with its tiny hind wing tails, similar to a hairstreak's. These tails serve the same function that they do in hairstreaks, deflecting bird attacks away from the real head and toward the thecla spot, with the tails acting as antennae. In contrast with some of the convergences mentioned earlier in the book, this is as far as the convergence between tailed blues and hairstreaks goes. The tails don't always help to identify the butterfly, however, as I discovered a few summers back. A

male Western Tailed Blue was flitting about in some greenery, stopping from time to time to sun itself, when it spied another blue. It flapped over to investigate, but the other butterfly seemed completely uninterested. That's when I saw a Flower Spider (*Misumena vatia*), tucked up between the purple flowers of the vetch where they were perched. Before the blue could realize its mistake, the spider reached out, grabbed it and had two blues for lunch instead of one. Not only had the spider used the flower for an ambush, it had also used the first blue as a decoy (albeit probably inadvertantly).

Also Called: *Everes amyntula*.

ID: a tailed blue with black wing margins on the topside and a white fringe; there can be one or two or three black spots without orange caps, near the tail base on the topside; underside is white, with indistinct spots; a single orange capped dark spot near the tail base underside. *Male:* mostly purplish blue on the topside. *Female:* mostly brown on the topside, with some blue.

Similar Species: *Eastern Tailed Blue* (p. 154): the only other tailed blue that has darker underwings, more distinct markings and a different pattern of spots near the tail base.

Caterpillar Food Plants: various bean-family plants (Fabaceae: *Vicia americana*, *Astragalus* spp., *Lathyrus* spp.).

Habitat & Flight Season: meadows, open forests and along rivers and streams; flies from late April to mid-June, and in July in more northern locales.

Spring Azure
Celastrina lucia

Wingspan: about 28–30 mm

♀

The Spring Azure is a classic butterfly. Everything about it is admirable. I think it has a great name, and it emerges early in the season (it hibernates as a pupa), when winter-weary naturalists are best able to appreciate a bright blue butterfly. Also, it is widespread enough that butterfly lovers all across the Northern Hemisphere can claim it as part of our shared natural heritage. What a shame, then, that it had to be split into four (or three, or six) North American species, all distinct from the original *C. argiolus*, the "Holly Blue" of Europe and northern Asia. It seems

Similar Species

Western Spring Azure

that, at first, lepidopterists were interested in the forms of the Spring Azure, and named *marginata* for the dark grey marginal markings on the underside, *lucia* for the central grey blob on the hind wing underside, and *violacea* for those with neither of these marks. Then, when new subspecies were identified, some of these names were reused in a more formal fashion, for example *Celastrina ladon lucia* (which is now considered a full species, *C. lucia*). Thus, in British Columbia, this butterfly can occur in its *lucia* form, *marginata* form or, rarely, in its *violacea* form. Let me leave you, then, with the words of the Comstocks, written in 1904: "What though the spring azure appears in Protean forms! The more incarnations of a butterfly so beautiful, the better" (*How to Know the Butterflies*, p. 254).

Also Called: Boreal Spring Azure; *Celastrina quesnellii*, *Celastrina ladon* or *Celastrina argiolus*.

ID: the sparsely spotted and often blotchy underwing pattern of azures is distinctive; small size; blue uppersides. *Male:* sky blue on the topside. *Female:* darker blue, with black wing tips and spots along the hind wing margin on the topside.

Similar Species: *Western Spring Azure* (p. 160): has darker, more purplish blue males, and generally lighter markings on the underside.

Caterpillar Food Plants: blueberries (Ericaceae: *Vaccinium* spp.), viburnum (Caprifoliaceae: *Viburnum* spp.), labrador tea (Ericaceae: *Ledum groenlandicum*), cherries (Rosaceae: *Prunus* spp.) and dogwood (Cornaceae: *Cornus* spp.).

Habitat & Flight Season: along rivers and streams, in meadows, open forests and shrubby areas; flies from early April to early June most years.

Western Spring Azure
Celastrina echo

Wingspan: about 28–30 mm

♀

This butterfly is only slightly different from the Spring Azure, so perhaps the best way to start its story is with the definition of the word "azure." The Western Spring Azure is more purplish than the Spring Azure, and is therefore more "azure" than its close relative. Usually, Western Spring Azures are of the form *violacea*, which was named for this violet colour (although it is frequently recognized by the absence of grey blotches or margins on the underwing rather than by any violet colour it may display). The *lucia* form also occurs in this species, but less so as one looks at examples from farther west toward the coast. The name *echo* is a reference to the similarity of this butterfly to an eastern

Similar Species

Spring Azure

relative with the name *pseudargiolus*. The name *pseudargiolus*, means "false *argiolus*," in reference to the first name that was applied to Spring Azures in North America, *argiolus*. So, we can conclude that the Western Spring Azure "echoes" the "false" Spring Azure, which in turn may be yet another species separate from the Spring Azure proper, *lucia*.

East of the Rocky Mountains, there is also a Summer Azure and a Dusky Azure, not to mention a Hops Azure in Colorado, an Atlantic Azure on the east coast, and an Appalachian Azure in the Appalachian Mountains Keeping track of new developments in the taxonomy of such butterflies is a good reason to join a lepidopterist group or list-serve.

Also Called: *Celastrina ladon echo, C. argiolus echo, C. ladon nigrescens* or *C. argiolus nigrescens*.

ID: the underwing is distinctively azure , but chalky white rather than grey with relatively faint, dark markings. *Male:* purplish blue on the topside. *Female:* wide, dark borders on blue upperside; spots near the hind wing margin are circled in grey.

Similar Species: *Spring Azure* (p. 158): Darker grey on the underside, and more often appears as the *marginata* or *lucia* forms.

Caterpillar Food Plants: ceanothus (Rhamnaceae: *Ceanothus* spp.), ocean spray (Rosaceae: *Holodiscus discolor*), possibly western spirea (Rosaceae: *Spirea douglassi*) and cherry (Rosaceae: *Prunus* spp.).

Habitat & Flight Season: along rivers and streams, in meadows, open forests and shrubby areas; flies in two broods; the first in April through early June, the second primarily in September.

Square-spotted Blue
Euphilotes battoides

Wingspan: about 20–28 mm

♂

The Square-spotted Blue is our one and only "buckwheat blue," the group that comprises the genus *Euphilotes*, and so named for the plants they feed on. The four North American species in this group all possess an orange-red aurora on the outer edge of the hind wing underside, but not the iridescent blue-green spots that the Melissa, Northern and Acmon Blues all share. The buckwheat blues spend most of their time on or near their caterpillar host plant, and the adults also take nectar from buckwheat blossoms. In fact, even the pupa remains on the stems of the host plant, where it overwinters. I suppose it is possible for a butterfly to spend its entire life on a single plant, although I doubt this happens very often. After all, they do fly. Speaking of which, it is possible that other buckwheat blues will eventually be discovered in British Columbia, but for the moment the Square-spotted has the province to itself. If you see one with smaller spots and a smaller aurora, it may well be the Rocky Mountain Dotted Blue (*Euphilotes ancilla*). Double check with other reference books to be sure, and if possible take a photograph (or a specimen if this is both legal and within the bounds of your own naturalist ethic) as a sample for what could be an interesting record.

In British Columbia and elsewhere, these butterflies are only now beginning to give up their secrets. Personally, I think that the dotted blues are the best example we have for a group that still needs to be extensively collected and studied by museum entomologists—without this process, we will never know how many species we are dealing with or where they are found.

ID: a wide, orange-red aurora band on the hind wing underside, with no adjacent blue or green spots; large, contrasting, angular dark spots on the underwings. *Male:* blue with wide, black borders, a slight aurora and some faint marginal spots on topside. *Female:* brown, with an aurora and adjacent dark spots on the topside.

Similar Species: the combination of aurora without iridescent spots make this species distinctive in B.C.

Caterpillar Food Plants: buckweat (Polygonaceae: *Eriogonum* spp.).

Habitat & Flight Season: found from deserts to open ridges, and in all the sunny places in between; flies from mid-May to early July, depending on altitude and temperature.

Arrowhead Blue

Glaucopsyche piasus

Wingspan: about 27–35 mm

A few years back, I was asked to participate in a survey of well-known butterfly watchers, in which we listed the species we wanted most to see. In it, I listed the Arrowhead Blue, and I'll be darned if I still haven't seen one alive! I'm not often in the right habitat, but when I am, it seems that other people see them just before or after I am there, so the Arrowhead Blue remains my "jinx butterfly." Actually, for me it should be "lynx butterfly" because I still haven't seen a lynx either. My interest in this butterfly comes not only from the fact that I have yet to see it, however. I also find it interesting because it is large, as blues go, and because it is closely related to the otherwise quite dissimilar Silvery Blue, and the now extinct Xerces Blue of California. Lepidopterists point out, as well, that the caterpillars of the Arrowhead Blue eat all parts of the host plant, not just the flowers and seeds, and in this sense they are also of interest. Ants do not seem to be as important to Arrowhead Blue caterpillars as they are to those of Silvery Blues, but this is not to say that Arrowhead Blue caterpillars are never tended by ants.

The name Piasus refers to a minor figure in Greek mythology who raped his daughter, after which she drowned him in a cask of wine. Clearly, this has nothing to do with the butterfly itself, but the tradition of using mythological names for butterflies was once very strong.

Also Called: *Phaedrotes piasus*.

ID: a large blue; underside has distinctive light-coloured, arrowhead-shaped marks. *Male:* bright blue with a narrow dark border and a checkered fringe on the topside. *Female:* dull blue with a wide, dark border and a checkered fringe on the topside.

Similar Species: the arrowheads make this species distinctive.

Caterpillar Food Plants: lupines (Fabaceae: *Lupinus* spp.), and possibly other bean family plants.

Habitat & Flight Season: pine forests, other open woodlands, sagebrush and meadows; flies from mid-May to mid July, depending on elevation.

Silvery Blue

Glaucopsyche lygdamus

Wingspan: about 22–30 mm

The Silvery Blue is only vaguely silvery in the west and, in fact, the blue colour is darker in the north (subspecies *G. l. columbia*) than in the south (subspecies *G. l. couperi*) of British Columbia. When the Comstocks addressed this species in 1904, very little was known of its biology, and in place of details they wrote, "All that we know is that it bears on its wings a blue found nowhere else in the world except in the pearly spectrum of the sea-shell…" (*How to Know the Butterflies*, p. 246). We now know that ants tend to the larvae and the pupae of these butterflies; the pupae often overwinter inside the nests of ants, which are fooled by the smell of the pupae into accepting them as their own. The larvae, when they are feeding on lupines, develop rather quickly, in less than a month. This is apparently a response to the production of toxic alkyloid chemicals by the plant where the leaves or flower parts have been chewed by caterpillars. It is essential for the caterpillar to complete development before the plant becomes inedible. Interestingly, when the larvae feed on flowers, they are white or purple in colour; when they feed on leaves, they are green. The pupae, on the other hand, are always pale, as are ant larvae and pupae, but in the darkness of an anthill, colour is of no consequence, of course.

Some individual Silvery Blues have very tiny dark spots on the underwing, or these may be absent altogether. This sort of variability is characteristic of some species of butterflies but not all, and is an interesting aspect of the study of butterflies in general.

ID: wings are grey on the underside; most individuals have an undulating line of black dots, circled with white, forming a "string of pearls" pattern. *Male:* blue above with a narrow black wing border. *Female:* blue at the wing bases fading to grey toward the outer edges, and with a faint cell bar above.

Similar Species: the underwing pattern is distinctive.

Caterpillar Food Plants: various perennial members of the bean family (Fabaceae).

Habitat & Flight Season: open areas where its food plant is found; begins its month-long flight season in April in warmer locales, and finishes in August in cooler places.

Northern Blue
Plebejus idas
Wingspan: about 20–30 mm

♂

The Northern Blue is one of three North American species in a closely related group, including the Anna's Blue and the Melissa's Blue. All three live in British Columbia, so this is as good a place as any to get to know them. They are not, however, easy to tell apart. In fact, the great novelist and lepidopterist Vladimir Nabokov devoted much of his butterfly time to this group of blues, and probably understood them as well as anyone before him and since. In any one place, you generally find only one of the three species, so identification is not a problem once you discover your favourite locations. They are all delightful butterflies, and the purplish blue hue of the male

Similar Species

Melissa's Blue

Northern Blue

can be recognized at a distance, with a bit of practice, and distinguished from the generally lighter blue colours of other species in the blue group. The females of this group are generally easy to recognize, with their orange auroras on the upper hind wings. This makes them my favourites among the female blues, but many female Northern Blues, such as the one illustrated on this page, have only a trace of orange on the upper surface. In British Columbia, the Northern Blue is generally classified into three subspecies, of which the most northern (*P. i. alaskensis*) has bluer females and smaller markings on the underwings. Some lepidopterists include Anna's Blue within the Northern Blue, as a subspecies, but here I will present it as a species in its own right.

Also Called: *Lycaeides idas, Plebeius idas, Plebejus argyrognomon* or *Plebeius argyrognomon*.

ID: grey underside; an orange-red aurora on both wings, with iridescent blue spots; pale spots with black centres at the centre of each wing. *Male:* purplish-blue on the topside with a narrow, dark border and a white fringe. *Female:* brown on the topside (except in the far north, where they are blue) with a hint of a dorsal aurora on the hind wing.

Similar Species: *Melissa's Blue* (p. 172): a slightly more extensive aurora on the topside and underside; lives in warmer areas at lower elevations, generally below 1000 m.

Caterpillar Food Plants: dwarf bilberry (Ericaceae: *Vaccinium caespitosum*).

Habitat & Flight Season: a variety of open meadows; flies in July and August.

Anna's Blue

Plebejus anna

Wingspan: 30–32 mm

If it weren't for the taxonomic confusion surrounding this butterfly, we would think of it as nothing more (or less) than a local colour variation of the Northern Blue. Many experts have done exactly that, including John C. Downey, who wrote in William Howe's great *Butterflies of North America* (1975) that the Anna's Blue interbreeds with the *ricei* subspecies of the Northern Blue, and thus they should be considered the same species. This has not convinced all specialists, however, and the fact that Anna's Blues feed on lupines is evidence, to some, that it is indeed something separate and distinct. As well, there was once another related

Similar Species

Northern Blue

butterfly on Vancouver Island that is now extinct. Lepidopterists Crispin Guppy and Jon Shepard have named this extinct butterfly *Lycaeides anna vancouverensis*, to distinguish it from the still-living *L. a. anna*. In technical terms, however, both of these butterflies are, or were, structurally identical to Northern Blues, even with respect to the form of their genitalia. And among butterfly taxonomists, genitalia counts for a lot. I wish I could give you a definitive pronouncement on this situation, but alas, I cannot. All I can tell you is that the Anna's and Northern Blues most certainly came from a common ancestor quite recently in the evolutionary past, and that the degree to which they have gone their own separate phylogenetic ways is slight. I can also tell you that "Anna" was yet another mystery woman in the life of W.H. Edwards, the 19th-century taxonomist. This butterfly came into the world of science shrouded in intrigue and has stayed that way.

Also Called: Anna Blue or part of the Northern Blue; *Lycaeides idas anna* or *Lycaeides anna*.

ID: a very pale underside; small underside spots; a bit of blue at the base of the hind wing underside. *Male:* blue on the topside, with a thin, black border and a white fringe. *Female:* brown above with orange aurora spots on the hind wing.

Similar Species: *Northern Blue* (p. 168): darker below, with larger spots, and no blue at the base of the hind wing underside.

Caterpillar Food Plants: lupines (Fabaceae: *Lupinus* spp.).

Habitat & Flight Season: open meadows at high altitudes; flies in July and August.

Melissa's Blue
Plebejus melissa
Wingspan: about 20–30 mm

♂

Named for another of W.H. Edwards' mystery women (either that or for honeybees, the Greek for which is "mellissa"), this is, in my opinion, the loveliest of all the *Lycaeides* blues. In British Columbia it is found only in the southern Interior, but to the east it is widespread across the prairies, where it replaces the Northern Blue south of the boreal forest. The Melissa's Blue has become famous in conservation circles because of one of its eastern subspecies, the Karner Blue (*P. m. samuelis*), originally named by Vladimir Nabokov. This butterfly has become increasingly rare in its native pine barrens habitat, and in fact has

Similar Species

Northern Blue

Anna's Blue

disappeared entirely from its Canadian range in southern Ontario. Various programs have been initiated to save the butterfly, and its habitat, in both Canada and the northeastern United States, with various degrees of success. While visiting a preserve in Albany, New York, for example, I found the butterfly much more common in a nearby borrow pit than in the carefully managed pine woods that had been set aside for the Karner Blue. Sigh... sometimes conservation is so embarassing! Clearly, the butterfly is simply a pawn in a broader struggle against development and "progress," and the real way to save the butterfly would be to bulldoze some sandy ground, let the lupines move in, and ignore the fact that scenic pine forests belong to a separate issue. And yes, some people consider the Karner Blue a separate species, just like the Anna's Blue. This sort of splitting seems inevitable when conservation motives enter into the mix, and a species is seen to have more value than a "mere" subspecies.

Also Called: Melissa Blue; *Lycaeides melissa* or *Plebeius melissa*.

ID: underside is pale grey with a pattern of dark spots, and an orange and iridescent blue aurora. *Male:* purplish blue on the topside, with a narrow dark border and a white fringe. *Female:* brown on the topside, with an orange aurora.

Similar Species: *Northern Blue* (p. 168) and *Anna's Blue* (p. 170): have smaller spots of all shapes and sizes.

Caterpillar Food Plants: various bean family plants (Fabaceae), including wild licorice (*Glycyrrhiza lepidota*) and lupines (*Lupinus* spp.).

Habitat & Flight Season: sagebrush areas and open pine forests; flies in two broods, the first from mid-May to early July, the second from late July to early September.

Greenish Blue

Plebejus saepiolus

Wingspan: about 22–30 mm

♂

Similar Species

Boisduval's Blue

In eastern North America, Greenish Blues actually do have a greenish hue that, when it can be seen, is near the wing bases. But here in western North America, they do not. Still, this is a familiar species to most naturalists, in part because it frequents disturbed, weedy places. The spot pattern on a Greenish Blue is quintessentially blue-ish, and echoes of this general pattern can be seen in the underwings of almost all our other species. Luckily, the only other species that closely resembles the Greenish Blue is the Boisduval's. Like the Anna's Blue, the Greenish once had a subspecies on Vancouver Island (*P. s. insulans*) that is now, apparently, extinct. These lost

Greenish Blue

butterflies are reminders of how terribly overdeveloped the Island now is, despite its "natural" feel. What a shame, right here in a part of Canada where people take such pride in their natural surroundings! The name *Plebeius*, by the way, means plebian, like a commoner or peasant. For this reason, I prefer the spelling *Plebeius* to the more commonly seen *Plebejus*, although the latter is now the standard. There is no letter J in either Latin or Greek, and that is the reason for this discrepancy. Greenish Blues are generally easy to identify, but worn individuals can give you pause as you consider which other species they might belong to.

Also Called: *Plebeius saepiolus*.

ID: underside is grey with a double row of spots, a cell bar and a few other spots at the base and leading edge of the hind wing; the cell bar can be visible on the topside; small orange thecla spot is present. *Male:* lighter on the underside, blue on the topside, with wide dark margins, dark spots near the margin and a white fringe. *Female:* browner on the underside, brown on the topside, with some blue scaling at the base of the wings; sometimes has some orange on the hind wing.

Similar Species: *Boisduval's Blue* (p. 176): has almost no cell spot on the topside, no thecla spot and reduced outer dark spots on the underside and larger spots on the forewing underside than on the hind.

Caterpillar Food Plants: introduced clovers (Fabaceae: *Trifolium* spp., but not *T. pratense*).

Habitat & Flight Season: fields and meadows, and often disturbed places, such as roadsides; flies from late May to July, and sometimes into August.

Boisduval's Blue

Plebejus icarioides

Wingspan: about 25–35 mm

This is another abundant, familiar type of blue, but it tends to live in isolated colonies so that it seems locally abundant, while seeming rarer elsewhere. Lupine-feeding butterflies are often like this. Speaking of which, the caterpillars seem to begin by feeding on the leaves of lupines, but switch to the flowers later in their lives. They overwinter about half way through their larval stage, which results in them finishing growth in the spring, and emerging later than the blues, which overwinter as pupae. The British Columbia populations of this species have been classified into three subspecies; one in the southeast, one at high elevations in the Interior and one on the Coast and the Island.

Interestingly, however, John C. Downey (in Howe's *Butterflies*

Similar Species

Greenish Blue

Sooty Hairstreak

Arctic Blue

of North America) reports that he once reared high-elevation caterpillars at low elevation and found that they developed into typical low-elevation butterflies. This sort of thing, which is more common than you might think among butterflies, is a good reason to take the "reality" of subspecies with a grain of salt. Certainly the differences among the butterflies are real, but this does not necessarily mean that they lie on different "evolutionary trajectories" as the current terminology is wont to say. The complex interactions of genes and environment are at the heart of all biology, and anyone who impatiently oversimplifies them will eventually find themselves humbled by the facts of the matter.

Also Called: Icarioides Blue; *Plebeius icarioides* or *Icaricia icarioides*.

ID: lightly coloured on the underside, with two rows of lightly coloured spots, with or without black centres on the hind wing. *Male:* lighter on the underside, blue on the topside with dark upper borders wider on the forewing. *Female:* browner on the underside, brown on the topside with blue wing bases and wide, dark margins.

Similar Species: *Greenish Blue* (p. 174), *Sooty Hairstreak* (p. 134) and *Arctic Blue* (p. 182): can have light spots on the hind wing underside that do not have dark centers.

Caterpillar Food Plants: lupines (Fabaceae: *Lupinus* spp.).

Habitat & Flight Season: grassy, sagebrush or forest meadow areas; flies from mid-May to August, depending on elevation.

178 BLUES • Family Lycaenidae | Subfamily Polyommatinae

Acmon Blue

Plebejus acmon

Wingspan: about 20–30 mm

This is one great looking blue. In fact, it's my favourite. The violet-blue upperwings with a dorsal aurora are simply stunning on a freshly emerged male. The name is simple and easy to memorize, but apparently refers to an annoying dwarf in Greek mythology who was in the habit of pestering Hercules. Sure, this butterfly is small, but I can't imagine any person or creature finding it annoying. In a general sense, the Acmon Blue is most similar to the Northern and its relatives, even though it is placed in the same

Similar Species

Melissa's Blue

Northern Blue

Anna's Blue

genus as the Boisduval's. This decision was made by Vladimir Nabokov, on the basis of similarities in genitalia. As many authors have commented, the taxonomic study of this butterfly may be far from over, and not only might the genus still change, the boundaries between Acmon Blues and their close relative the Lupine Blue (*Icaricia acmon*) still need to be clarified as well.

This butterfly is rarely common, and usually turns up one at a time, here and there. Why one species should be rare while closely related species are common is one of those fundamental questions in biology for which we still have only fuzzy, partially satisfying answers.

Also Called: *Plebeius acmon* or *Icaricia acmon*.

ID: B.C.'s only blue with a broad orange aurora on the upper hind wing and not the forewing, in both sexes. *Male:* purplish blue on the topside; orange-pink, upper aurora with dark-spotted centres. *Females:* brown topside, with blue wing bases and an upper aurora and a cell spot.

Similar Species: *Melissa's Blue* (p. 172): males have no orange on topside, females have more orange on topside. *Northern Blue* (p. 168) and *Anna's Blue* (p. 170) have smaller underwing spots.

Caterpillar Food Plants: native bean family plants (Fabaceae), as well as shrubby Polygonaceae, such as buckwheat (*Eriogonum* spp.).

Habitat & Flight Season: a variety of open habitats, including meadows, sagebrush, open pine forests; flies from mid-May at low elevations to mid-August up high, with a partial second brood in the Okanagan.

Cranberry Blue
Plebejus optilete
Wingspan: about 22–28 mm

♂

The blues are a north-temperate group of butterflies, and the Cranberry Blue is one of the most northern of the lot. It is found in both the New and the Old Worlds, and there is only one species in the genus. In North America, the subspecies *yukona* is found in, and named for, the Yukon Territory. The name of the genus, by the way, is derived from the genus name of the food plant, *Vaccinium*, the blueberries, cranberries and lingonberries. When you find a Cranberry Blue, you generally find just that—one Cranberry Blue. They are not the sort of butterfly, at least in my experience, that occurs in great swarms. One here, one there, flitting about in peatlands, is the usual

Similar Species

Greenish Blue

scenario. And sooner or later you will find that they are quite easy to recognize, just from the general "look" of their underwings. I think of Cranberry Blues as small, dark, contrasting and solitary. I also think of them as a "good find," and a day with a Cranberry Blue is always worthy of an entry in my field notes.

The Cranberry Blue reminds me very much of a prairie species from east of the Rocky Mountains, the Shasta Blue. But while the Cranberry Blue is a creature of boggy lowlands, the Shasta flies high on valley tops and bluffs. Shasta Blues have not yet been found in British Columbia.

Also Called: Yukon Blue; *Plebeius optilete* or *Vacciniina optilete*.

ID: a blue with the typical spot pattern on the underside, although the spots are large and distinct, as well as an orange spot in the thecla position. *Male:* deep blue-purple on the topside with no dark borders. *Female:* brown with a bit of blue on the topside, and wide, dark borders.

Similar Species: *Greenish Blue* (p. 174): similar, but illustration comparisons should prevent confusion.

Caterpillar Food Plants: blueberry and cranberry (Ericaceae: *Vaccinium* spp.).

Habitat & Flight Season: bogs and wet tundra; flies in July and the first half of August.

Arctic Blue

Plebejus glandon

Wingspan: about 25–28 mm

Lepidopterist John C. Downey wrote that "this butterfly has an 'Arctic' look about it, particularly its small size and dull colors" (*Butterflies of North America*, p. 349). My only response to this obviously circular line of reasoning is that polar bears, musk oxen and brightly coloured arctic wildflowers must not know about this rule! Not to put Downey down—if one forms one's opinions about the entire Arctic on the basis of a single butterfly (and OK, I'll admit there are some other small, dull arctic butterfly species), this says quite a

Similar Species

Boisduval's Blue

Sooty Hairstreak

bit about how strongly butterflies can influence our views about the world. More to the point, Arctic Blues are amazingly easy to recognize on the wing because they really do flutter when they fly, and their colours are sufficiently grey, distinguishing them from the other, more glittering blues. Many authors point out that Arctic Blues typically fly low to the ground, but this is typical of many blues, and doesn't really help in the field. In the Peace River grasslands of British Columbia and Alberta, the Arctic Blues are considered by some to form a distinct subspecies, *Plebejus glandon lacustris*. This taxon, if indeed it is distinct, is of conservation concern in the province.

Also Called: Rustic Blue; *Agriades aquilo*, *Plebius aquilo* or *Agriades glandon*.

ID: underside has a band of white through the hind wing, and sometimes an orange and blue thecla spot; upperside cell bar with a light outline and small, dark spots outlined with lighter halos on the hind wing. *Male:* grey-blue topside, grading to grey on the wing margins. *Female:* brown topside; topside has some blue in the northern parts of this butterfly's range.

Similar Species: Among the blues, this is one of the easiest to recognize. *Boisduval's Blue* (p. 176) and *Sooty Hairstreak* (p. 134): similar on the underside.

Caterpillar Food Plants: a variety of plants, including saxifrage (Saxifragaceae: *Saxifraga* spp.), and possibly primrose (Primulaceae), locoweed and rock jasmine (Fabaceae: *Oxytropis* and *Androsace*).

Habitat & Flight Season: in the south of the province, Arctic Blues are found only at high elevations, at or above treeline; farther north, they are found in a variety of open areas; flies mainly in June and July, but a bit later in the alpine.

Metalmarks
(Family Riodinidae)

Metalmarks

Metalmarks are small-to medium-sized butterflies that reach their greatest diversity in the American tropics. There, they are incredibly varied in appearance: some look like tiny swallowtails, some like moths, and others like just about any other sort of butterfly you care to name. In the southern and eastern United States, it is common for metalmarks to have small, iridescent patches in their wing patterns, and these are the "metal marks" for which the group is named. Only one species reaches western Canada, and it (not surprisingly) has no metal marks and looks, at first glance, more like a checkerspot than anything else.

The metalmarks are treated here as a family, but there has been much debate about whether or not they should be grouped with the gossamer-wings as a subfamily instead. At the very least, it seems clear that the two groups are closely related.

Mormon Metalmark, photographed in Sonora, Mexico

Mormon Metalmark

Apodemia mormo

Wingspan: 25–35 mm

♂

This butterfly has a very broad geographic range, and, in fact, the first time I encountered it myself, I was in Sonora, Mexico. It is also quite variable throughout its range (most of the western half of the U.S. and some parts of northwestern Mexico), and it may well deserve to be split into a number of separate species. Mormon Metalmarks are closely tied to their buckwheat host plants, and are butterflies of late summer and fall. In British Columbia, they are now known only to inhabit the southernmost Okanagan, near Keremeos. Here, they are considered endangered, but I have never been comfortable using such a strong word for any animal that is still hugely abundant farther south. It is considered endangered here only because the political border between Canada

Similar Species

Variable Checkerspot

and the United States happens to coincide with the northern extent of its range. Still, it is likely that we will lose the Mormon Metalmark from the British Columbia fauna if its habitats are not protected. Habitat change is now widespread in the southern Okanagan, and other species of plants and animals have in fact disappeared already.

One possible solution might be for gardeners in the southern Okanagan to plant buckwheat foodplants and not spray them with pesticides. Near by, they could provide fall-blooming nectar flowers such as asters, so the adults have food. In other places, this sort of thing has reversed the decline of some butterflies, and if you want to look up a particular example, the Atala in southern Florida is the one that springs most quickly to mind.

ID: a small to medium-sized brown, or orange and brown butterfly with white speckles; distinctive overall.

Similar Species: some checkerspots (pp. 242–50) are similar, but not on close examination.

Caterpillar Food Plants: snow buckwheat (Polygonaceae: *Eriogonum niveum*).

Habitat & Flight Season: flies in open, dry areas from mid-August through September.

Brush-footed Butterflies
(Family Nymphalidae)

Brush-footed butterflies have four walking legs, not six, and the front legs have been reduced over the course of evolution to form a pair of small "brushes." In that sense, the family is easy to define, but to the casual naturalist, membership in the group is not as easy to recognize at a glance as it is in any of the preceding groups. Brush-foots are generally prone to gliding between bursts of flapping flight, and they are, more often than not, marked in browns and oranges. Beyond these simple generalizations, however, the variation within the family is immense.

There are two prominent subfamilies within the Nymphalidae that are often given their own family status, and this makes characterization of the group difficult as well. These are the satyrs and the milkweed butterflies, and both have representatives in British Columbia. The remaining brush-footed butterflies fall into four other subfamilies, and it does help to learn the subfamilies as you learn the species because doing so categorizes the diversity into manageable chunks. As in most butterfly books, we begin the brushfoots with the fritillaries.

Northwestern Fritillary

Fritillaries
(Subfamily Heliconiinae)

The first subfamily within the brushfoots is the fritillaries, and it is only recently that lepidopterists have formally acknowledged the close relationship between the familiar orange and black, temperate fritillaries proper, and the long-winged, tropical heliconians. In the United States, the heliconians include such species as the Julia and the Zebra Longwing, as well as the Gulf Fritillary, a butterfly roughly half-way between a typical fritillary and a typical heliconian. None of these beauties lives in British Columbia.

Our fritillaries fall into three genera, and thus three natural groups. The Variegated Fritillary is a visitor from the South, and in British Columbia it is alone in its genus. The greater fritillaries are typified by

Introduction to the Fritillaries 191

medium to large size, and silver spots on the hind wing underside. This group is notorious for being difficult to sort into species, but it is easy to recognize a greater fritillary as such. It is also worth pointing out that, although the species are tough to tell apart, there has been less confusion among them historically than in many other butterfly groups.

The lesser fritillaries are less similar to one another than are the greater fritillaries, both with respect to underwing pattern and larval food plants. Most lesser fritillary caterpillars feed on violets, as do caterpillars of greater fritillaries, but those that live in colder alpine or arctic habitats feed on other vegetation.

If the truth should be known, most beginners dislike fritillaries because of the identification challenges, but as with sulphurs, once you get the hang of them, they can quickly become favourites. After all, who wants butterflies to be made simple?

Don't feel bad if you can't identify this fritillary. I think it's an Aphrodite, but I can't be sure.

Variegated Fritillary

Euptoeita claudia

Wingspan: about 45–65 mm

♀ ♂

The Variegated Fritillary is not known to breed in our area, but instead it emigrates from the south each year, sometimes in great numbers. Individuals vary quite a bit in size and in colour, but because this is a wandering species, there are no regional subspecies. Like people, it seems to do well just about anywhere, and as you get to know it, you may come to feel a bit of this kinship yourself. There is a tendency among naturalists to be critical of ecological generalists, but this is mostly a matter of psychological territoriality on our part. As John H. and Anna Comstock wrote: "not only with a mosaic of blossoming weed does Nature deck her waste

Similar Species

Callippe Fritillary

places; lest the flowers fade she scatters there many-hued butterflies; and by these as well as the blossoms she tells us plainly that she has no waster places where she has at hand water-power and sunshine-power to help her manufacture life and color. Above such lands neglected by man, the variegated fritillary hovers on golden red wings or rests basking in the sun on the sands of drought-wasted streams" (*How to Know the Butterflies*, pp. 109–110).

ID: a fritillary with a unique pattern of light and dark orange, brown and grey; sexes are similar.

Similar Species: greater fritillaries are somewhat similar, but only at a distance.

Caterpillar Food Plants: not known to breed in B.C., but feeds on a variety of plants elsewhere, including violets (Violaceae: *Viola* spp.).

Habitat & Flight Season: found in a variety of habitats, usually late in the season, although records exist for May, June, July, August and September.

Great Spangled Fritillary
Speyeria cybele

Wingspan: about 65–90 mm

T he Great Spangled is the flagship fritillary of North America, and in eastern North America, it is one of only three common species in its genus, and therefore very easy to recognize. There are two subspecies in British Columbia. In the Peace River region, one finds the *pseudocarpenteri* race, which is similar to the classic eastern forms of the species, with a lot of rich orange in the colour pattern. In the southern Interior, the enigmatic *leto* race is found. Leto Fritillaries are strongly sexually dimorphic, with females that are very dark on the wing bases, and very

Similar Species

Aphrodite Fritillary

pale in ground colour. As with most greater fritillaries, the females are less frequently encountered than the patrolling males. Some authors have treated *leto* as a separate species, but I follow the majority view, that *leto* belongs to the same species as the other subspecies of the Great Spangled Fritillary. The main thing to remember is that, in typical fashion, this species is much more complicated in its biology and evolutionary history in western North America than it is in eastern North America. It was originally described from New York City, but that doesn't mean that westerners aren't the people who know it best.

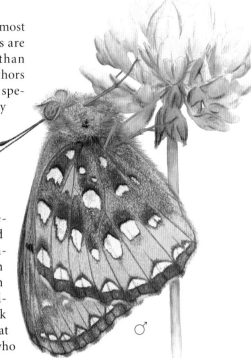

Also Called: Leto Fritillary; *Speyeria leto*.

ID: a large fritillary with dark wing bases on the topside; an especially wide, pale band between the outer and inner rows of silver spots on the underside hind wing; eyes are brown when alive (but this difference is not apparent in dead specimens, in which the eyes usually turn dark brown even in butterflies that had grey eyes while alive); female, especially in the south, is somewhat larger, and much paler than the male on the topside.

Similar Species: other greater fritillaries are generally smaller and have a more narrow pale band on the hing wing underside.

Caterpillar Food Plants: violets (Violaceae: *Viola* spp.).

Habitat & Flight Season: fields and clearings; flies in June, July and August.

Callippe Fritillary
Speyeria callippe
Wingspan: about 60–65 mm

Similar Species

Mormon Fritillary

I find that the way a Callippe Fritillary looks starts to make sense long before you become familiar with the finer points of recognizing this species. The green scaling on the hind wing underside can be quite beautiful and certainly adds an easy field mark to the pattern. But green underwings are not entirely characteristic of Callippes: those in the Chilcotin region are more brownish on the underside, and have been named by lepidopterists Crispin Guppy and Jon Shepard as a separate subspecies. All Callippes hilltop, however, and they even migrate upslope in the heat of summer, returning to lower elevations when the weather begins to cool. With a long flight season, they are often the first greater fritillaries seen each year. In the older literature, it was not clear whether our Callippe Fritillaries should be called Nevada Fritillaries (*S. nevadensis*) instead, but this seems to have been resolved, at least for the moment. I personally think of this butterfly as a prairie species, since it is abundant east of the Rocky Mountains and out onto the grasslands of the southern prairie provinces. It was a surprise to me to encounter it above treeline in the Rockies of Alberta, but this experience prepared me well for appreciating its ecology west of the Continental Divide in British Columbia.

ID: a greater fritillary with a greenish tinge on the underside and the hind wing silver spots showing through as light orange areas on the topside; sexes are similar.

Similar Species: *Mormon Fritillary* (p. 207): can be green on the underside, but is much smaller. None of our other greater fritillaries show the same pattern of pale orange spots on the upperwing surfaces; sexes are similar.

Caterpillar Food Plants: violets (Violaceae: *Viola* spp.).

Habitat & Flight Season: open areas; flies from May through August at a variety of elevations.

Zerene Fritillary

Speyeria zerene

Wingspan: about 55–65 mm

♂ ♀

Like the Callippe, this butterfly will migrate upslope to the hilltops, and then return to lower areas to lay eggs. Zerene Fritillaries are paler in drier areas, but with four subspecies in British Columbia that are quite different, they are very hard to recognize, except by default once all

Similar Species

Great Spangled Fritillary   Aphrodite Fritillary   Callippe Fritillary   Mormon Fritillary

other species have been eliminated. In the far northwest corner of the province, they are sooty on the topside and reddish on the underside, with small silver spots—quite a distinct look. On Vancouver Island and the Lower Mainland they are very dark on the topside, with a dark red-brown underwing ground colour, like the Atlantis Fritillary. Throughout the rest of the southern portions of the province, this butterfly is lighter in colour, with larger silver spots. They might be confused with Northwestern Fritillaries, with which they share grey eyes when alive, but the latter possess smaller silver spots, and thinner, dark markings on their topside.

♂ ♀

ID: a large, strongly marked greater fritillary, but not easy to characterize as a species; sexes are similar.

Similar Species: this species is larger than all but the *Great Spangled Fritillary* (p. 194) and *Aphrodite Fritillary* (p. 200). *Callippe Fritillary* (p. 196) and *Mormon Fritillary* (p. 207): both are green on the underside; sexes are similar.

Caterpillar Food Plants: violets (Violaceae: *Viola* spp.).

Habitat & Flight Season: prefers meadows and open habitats, including sagebrush areas; flies mostly in August, but in some areas as early as late May.

Aphrodite Fritillary
Speyeria aphrodite
Wingspan: about 55–65 mm

♂ ♀

Similar Species

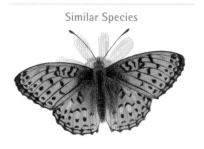

Northwestern Fritillary

Like the Great Spangled and the Atlantis, this is also a fritillary of eastern North America. There, it is easy to recognize, but here in western Canada, as usual, it has caused confusion, especially with respect to the Northwestern Fritillary. I used to remember it by the fact that the male Aphrodite does not have thickened wing veins for "sex scaling" and, because Aphrodite was the Goddess of Love, why would these butterflies need it? Now, thanks to the American author Jeff Glassberg, we know that the eye colour of living Aphrodite Fritillaries is brown while that of the

Northwestern is grey. This helps immensely but be sure, however, to examine fritillaries from the side when judging eye colour—from the top they all look brown. The wing pattern differences between the species are subtle, to say the least. The most useful characteristic for identifying the Aphrodite has been that, on the hind wing below, the small brown spot just out from the hindmost large silver spot in the inner band is "haloed" in light orange thinly surrounded by darker brown. The Aphrodite Fritillary always looks clean and bright, and it is a lovely butterfly. In eastern North America, it is a bit larger and darker orange, but it still contrasts nicely with the other two common eastern species, the Great Spangled and the Atlantis.

ID: a medium to large, bright orange greater fritillary, with brown eyes when alive, and a reddish brown hind wing ground colour; sexes are similar.

Similar Species: *Northwestern Fritillary* (p. 204): grey eyes, but is otherwise almost identical most of the time.

Caterpillar Food Plants: violets (Violaceae: *Viola* spp.).

Habitat & Flight Season: meadows, clearings and open areas; flies from mid-June to mid-September.

Atlantis Fritillary

Speyeria atlantis

Wingspan: about 55–65 mm

♂ ♀

The Altantis is the third of the three common eastern greater fritillaries, along with the Great Spangled and the Aphrodite. Because they are relatively easy to separate in eastern Canada, many people mistakenly assume this will be the case here as well. In British Columbia, there is only one subspecies of Atlantis (*S. a. hollandi*), and luckily it is relatively easy to recognize. However, some authors do not consider the Northwestern Fritillary to be a species separate from the Atlantis, and herein we encounter our most difficult taxonomic problem in the greater fritillary group. As far as we

Similar Species

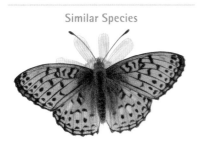

Northwestern Fritillary

can tell, the two "species" can interbreed south of B.C., in the Pacific Northwest, but not in western Canada, making the boundary between them fuzzy, at best. If the Northwestern Fritillary is considered a separate species (as it is here), the confusing aspects of the problem involve only the Northwestern, not the Atlantis proper, and for that I hope you are at least a little bit grateful. The literature on the subject occasionally refers to the disturbing possibility that both the Northwestern and the Atlantis might include individuals that are indistinguishable from the other species. And by the way, did you know that all the greater fritillaries in North America used to be called *Argynnis*, not *Speyeria*, and that lepidopterist Thomas Simonsen is now arguing quite persuasively that we should return to the original name (shared with the larger fritillaries in Europe and elsewhere)?

ID: a medium to large greater fritillary; dark orange and brown and solid dark wing borders on the topside; chocolate brown around the silver spots on the underside; sexes are similar.

Similar Species: *Northwestern Fritillary* (p. 204): generally lighter in colour.

Caterpillar Food Plants: violets (Violaceae: *Viola* spp.).

Habitat & Flight Season: mainly in forest clearings; flies in June, July and August.

Northwestern Fritillary
Speyeria hesperis
Wingspan: about 45–60 mm

Northwestern Fritillaries generally occur in drier habitats than Atlantis Fritillaries. As well, it is good to know that there are two distinct subspecies of the Northwestern Fritillary found in

Similar Species

Aphrodite Fritillary

Atlantis Fritillary

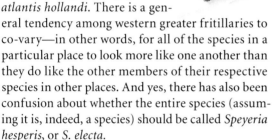

British Columbia. *Speyeria hesperis helena* (=*S. h. lais*) is a lot like the Aphrodite Fritillary and lives in the Peace Country as well as in aspen forests to the east, in Alberta. The other is *Speyeria hesperis beani* (which some think deserves a separate name in British Columbia because it is larger than the *beani* from farther east) and looks much more like the Atlantis Fritillary, or at least more like the subspecies *Speyeria atlantis hollandi*. There is a general tendency among western greater fritillaries to co-vary—in other words, for all of the species in a particular place to look more like one another than they do like the other members of their respective species in other places. And yes, there has also been confusion about whether the entire species (assuming it is, indeed, a species) should be called *Speyeria hesperis*, or *S. electa*.

Also Called: Atlantis Fritillary, in part, Northwestern Silverspot or Electa Fritillary; *Speyeria electa* or *Speyeria atlantis*, in part.

ID: a medium-sized greater fritillary with grey eyes when alive; in the northeastern portions of the province (including the Peace Country), the upperwing has a double-line border; silver spots on the hind wing have a reddish brown outline; in most of the rest of the province, the border is black and the hind wing is darker reddish brown; sexes are similar.

Similar Species: in the Peace Country, the brown-eyed *Aphrodite Fritillary* (p. 200) is most similar. Elsewhere, the *Atlantis Fritillary* (p. 202) has a darker brown colour around the silver spots.

Caterpillar Food Plants: violets (Violaceae: *Viola* spp.).

Habitat & Flight Season: clearings in dry forests and open grasslands; flies mainly in June, July and August.

Hydaspe Fritillary

Speyeria hydaspe

Wingspan: about 45–60 mm

The Hydaspe Fritillary is one of the easier members of the group in British Columbia to identify. It is also remarkably attractive, and the darkened hues of both the upperwing and underwing give it a rich look that other fritillaries never quite match. A freshly emerged individual can be especially impressive. I find it interesting, however, that W.J. Holland remarked that "It is quite impossible to distinguish typical specimens of *A. zerene* Boisduval from typical specimens of *A. hydaspe* named by the same author seventeen years later" (*The Butterfly Book*, p. 95). Presumably he was referring to these two species when they appear south of the U.S. border, proving, once again, that two fritillaries that are easy to recognize in one area may be downright impossible to identify elsewhere. The "*A.*" W.J. Holland refers to is for *Argynnis*, an old name for greater fritillaries that is still used for similar-looking butterfly species found in Europe.

Also Called: Lavender Fritillary.

ID: this greater fritillary has silver spots (that are often not silver) surrounded by a purplish colour; the pale area between the inner and outer row of silver spots is very narrow or absent; dark hind wing underside; sexes are similar.

Similar Species: the overall dark hind wing underside distinguishes this species.

Caterpillar Food Plants: violets (Violaceae: *Viola* spp.).

Habitat & Flight Season: meadows, especially in the mountains; flies from June through August.

Mormon Fritillary

Speyeria mormonia

Wingspan: about 40–50 mm

To finish the greater fritillaries, it is nice to encounter another easy-to-identify species. The Mormon Fritillary is not only smaller than other *Speyeria*, but it's also built differently, with a noticeably larger body in relationship to its wing size. It's tough to mistake it for any other greater fritillary, but in the field you should be aware that at a distance it can look a lot like a lesser fritillary. I personally enjoy this species a great deal because each individual I meet usually presents some sort of surprise. Either I am not expecting the species in a particular habitat, or I am intrigued by the look of the individual in question. Some are light in colour, some are dark. Some have brown eyes, some have grey. Some have silver spots, some have pale orange spots. And so on! This is a great butterfly for those who are interested in variation, and few other species in British Columbia can match it in this regard.

ID: a small, but highly variable greater fritillary, often with unsilvered, cream-coloured underwing spots; eyes can be brown or grey while alive; sexes are similar.

Similar Species: the small size of the Mormon Fritillary distinguishes it from all of its close relatives.

Caterpillar Food Plants: violets (Violaceae: *Viola* spp.).

Habitat & Flight Season: low grasslands up to the alpine; flies mainly in July and August.

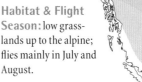

Mountain Fritillary
Boloria alaskensis
Wingspan: about 30–40 mm

The Mountain Fritillary is a reminder of the ice age. This species' range extends across the Canadian Arctic and down along the Rockies into the area of the "bend" in the British Columbia/Alberta border. It is closely related to a Eurasian species (*B. napaea*), and for a long time they were considered one and the same. There is also an isolated population of Mountain Fritillaries in the mountains of Wyoming that I predict will eventually be given its own name, just because it is so far removed from its polar relatives. Mountain Fritillaries are typical of the lesser fritillary

Similar Species

Freija Fritillary

genus *Boloria*, but they are also unique in that this species is the only member of our fauna that is also a member of the subgenus *Boloria* (*Boloria*). As such, it has always remained a "true *Boloria*" even when other members of the genus have been called *Clossiana*. The closest relative of the Mountain Fritillary in British Columbia is the Bog Fritillary, which is truly surprising given the difference in their wing patterns. And be sure to notice that although the lesser fritillaries differ somewhat in their topside patterns, the underwings are where the really diagnostic features lie.

Also Called: *Boloria napaea*.

ID: a lesser fritillary with a slightly angled hind wing margin; unique underwing pattern includes a light area midway along the outer margin of the hind wing; female is darker and less orange than the male.

Similar Species: *Freija Fritillary* (p. 224): similar, but compare the underwing patterns closely.

Caterpillar Food Plants: unknown, although some authors suggest blueberries and cranberries (Ericaceae: *Vaccinium* spp.) while others report alpine bistort (Polygonaceae: *Polygonum viviparum*).

Habitat & Flight Season: clearings and streamsides near timberline; flies in July and August.

Bog Fritillary

Boloria eunomia

Wingspan: about 40 mm

♀ ♂

The Bog Fritillary is a butterfly of cool places—the boreal forest and the high elevations of the Rockies. It is, therefore, one of the northern butterflies we should take greatest pride in as Canadians. When you find the Bog Fritillary, you are usually in an interesting, wild place at the time. Up close, the underwing pattern is lovely and not difficult to identify once you have become familiar with the butterfly. Despite their name, Bog Fritillaries are not particularly fond of bogs, except apparently in eastern North America, where the species first received its English name.

Similar Species

Silver-bordered Fritillary

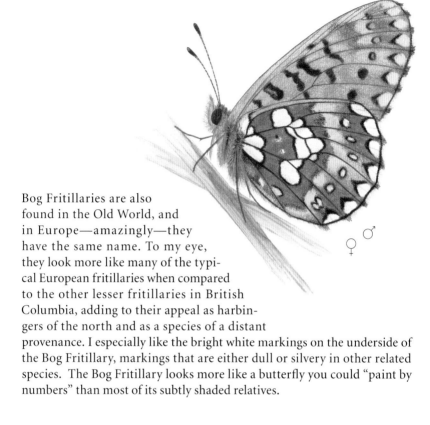

Bog Fritillaries are also found in the Old World, and in Europe—amazingly—they have the same name. To my eye, they look more like many of the typical European fritillaries when compared to the other lesser fritillaries in British Columbia, adding to their appeal as harbingers of the north and as a species of a distant provenance. I especially like the bright white markings on the underside of the Bog Fritillary, markings that are either dull or silvery in other related species. The Bog Fritillary looks more like a butterfly you could "paint by numbers" than most of its subtly shaded relatives.

Also Called: *Clossiana eunomia, Proclossiana eunomia.*

ID: white spots on the underwing, including a post-median row of circular dots; sexes are similar.

Similar Species: *Silver-bordered Fritillary* (p. 212): similar, but its light spots are generally silver, not white, and the row of white dots is replaced by black dots.

Caterpillar Food Plants: the caterpillar has not been discovered, at least not in North America, but probably feeds on violets (Violaceae: *Viola* spp.).

Habitat & Flight Season: moist meadows, and moist areas near streams, lakes and peatlands; generally flies from mid-June to early July.

Silver-bordered Fritillary

Boloria selene

Wingspan: about 40–45 mm

Similar Species

Bog Fritillary

The image of a Silver-bordered Fritillary sunning with its wings spread is one of my favourite butterfly sights. The upperwings are crisply patterned, without much sootiness near the body, which helps to make the butterfly recognizable from the top. The underwing, however, is the real "selling point" for this species. I find it interesting to think about the evolution of silver spots among members of the fritillary subfamily. Did the Silver-bordered Fritillary, and the greater fritillaries, inherit this feature from a silver-spotted common

ancestor? Or did it arise independently in the two groups? Intriguingly, the Variegated Fritillary genus, *Euptoeita*, shows no silver spots, but the Gulf Fritillary genus, *Agraulis*, does (this is a butterfly of the American south). My friend and colleague, Thomas Simonsen of Denmark, studies the relationships among fritillaries, and when I asked him this question, he found it so interesting that he promised to embark on a study to find the answer. So far, his answer is that it looks as if silver spots have arisen many times in the fritillary subfamily, and that silver colour is simply extra-bright white, reflecting as much of the visible light spectrum back as possible, but not flat enough to serve as a mirror.

Also Called: *Clossiana selene*.

ID: silver spots on the underside of the hind wings set this species apart from all other lesser fritillaries; dark-coloured mutants are more common than among other related species; sexes are similar.

Similar Species: *Bog Fritillary* (p. 210).

Caterpillar Food Plants: violets (Violaceae: *Viola* spp.).

Habitat & Flight Season: found in association with peatlands in B.C., although this is not always the case to the east of the Rockies; flies from late May to early September in the lower Okanagan Valley, where there are two generations (with a partial third) each year; farther north, there is only one generation per year, with adult numbers peaking in July.

Meadow Fritillary
Boloria bellona
Wingspan: about 45 mm

The aspen meadow habitat of the Meadow Fritillary is more abundant on the other side of the Rockies, giving this species a bit of an eastern flavour. By contrast, in extreme eastern North America,

Similar Species

Frigga Fritillary

Western Meadow Fritillary

Meadow Fritillary

this is one of only a few lesser fritillaries, making it a symbol of the West for many eastern butterfly fans. In British Columbia, we tend to think of the Western Meadow Fritillary "replacing" the Meadow Fritillary proper on this side of the Rockies, and it would be interesting to know if the two species came from a common ancestor that evolved in different ways on either side of the Continental Divide. The two species have been considered very close relatives, so perhaps this is precisely what happened. The Meadow Fritillary is a good-looking butterfly, and one that is easily recognized.

Also Called: *Clossiana bellona* or *Boloria toddi*.

ID: the underwing pattern is purplish with rounded rather than jagged markings; tip of the forewing is angular, as if cut by scissors; sexes are similar.

Similar Species: the forewing tip is a superb field mark, and distinguishes this species from the *Frigga Fritillary* (p. 216) and *Western Meadow Fritillary* (p. 220).

Caterpillar Food Plants: violets (Violaceae: *Viola* spp., especially *V. canadensis*).

Habitat & Flight Season: meadows in aspen forests; late May to late August in the south (two broods), but mainly in July in the North (one brood).

Frigga Fritillary
Boloria frigga
Wingspan: about 40 mm

Most of the butterfly watchers I know confuse this species, not with those that look most like it, but with the Freija Fritillary, because of the similar name. Both were named in Europe, by Carl P. Thunberg, after figures in Norse mythology, because

Similar Species

Meadow Fritillary

Western Meadow Fritillary

of their association with the north. B.C. lepidopterists Crispin Guppy and Jon Shepard point out that the Freija, Frigga, Bog and Arctic Fritillaries form a guild of boreal and high-altitude species that usually occur in the same areas, but at slightly different times during the season. Freija is the earliest to emerge, when it is still "frei-zing," and often there is still some snow on the ground. My mnemonic goes on to point out that only when Frigga emerges does it become "friggin'" confusing. Of course, the real way to remember all of this is to get to know the butterflies better. I can't help mentioning how cute the rhyming name of our British Columbia subspecies can be: *B. frigga saga*. Saga, of course, means just that (saga or story), and you might find, as I do, that you are humming "the frigga saga saga," "the frigga saga saga" to yourself as you hike through boreal meadows in search of these butterflies.

Also Called: *Clossiana frigga*.

ID: the bubbly patterned hind wing underside has a large white area near the base of the wing; sexes are similar.

Similar Species: *Meadow Fritillary* (p. 214): angular front wing tip. *Western Meadow Fritillary* (p. 220): lacks the big white blotch.

Caterpillar Food Plants: probably willows (Salicaceae: *Salix* spp.), but possibly blackberries and their relatives (Rosaceae: *Rubus* spp.).

Habitat & Flight Season: moist, open areas, similar to the Bog Fritillary; flies from late May to late July.

Dingy Fritillary

Boloria improba
Wingspan: about 35 mm

The specific epithet *improba* probably means something like "of lesser rank," and this butterfly was originally described as a subspecies of the Frigga Fritillary, of which it is clearly the smaller and less attractive. In turn, another butterfly that was originally described as a full species, the Uncompagre Fritillary (*Boloria acrocnema*) is now considered a subspecies of the Dingy Fritillary. I consider them all to be full species, despite the fact that they have

Similar Species

Frigga Fritillary

Beringian Fritillary

Alberta Fritillary

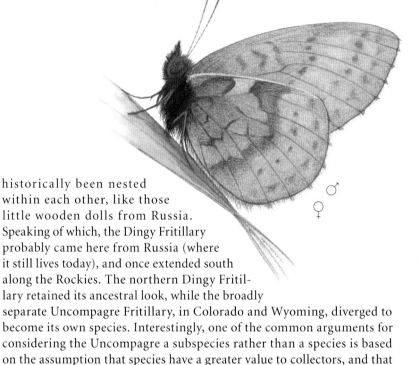

historically been nested within each other, like those little wooden dolls from Russia. Speaking of which, the Dingy Fritillary probably came here from Russia (where it still lives today), and once extended south along the Rockies. The northern Dingy Fritillary retained its ancestral look, while the broadly separate Uncompagre Fritillary, in Colorado and Wyoming, diverged to become its own species. Interestingly, one of the common arguments for considering the Uncompagre a subspecies rather than a species is based on the assumption that species have a greater value to collectors, and that collectors might then harm this butterfly's populations. I won't say this is impossible, but I do resent this sort of political interference with taxonomy. I suspect that over-collecting is much more likely among more colourful butterflies, and that full species rank also brings with it greater conservation concern.

Also Called: *Clossiana improba*.

ID: the smallest lesser fritillary; has a more washed-out underside and a sootier-looking upperside than the Frigga Fritillary; sexes are similar.

Similar Species: *Frigga Fritillary* (p. 216): very similar in pattern, but larger. *Beringian Fritillary* (p. 225) and *Alberta Fritillary* (p. 222): similarly "dingy" on the upperside and live in the same alpine habitats; underwing patterns are quite distinctive.

Caterpillar Food Plants: dwarf willows (Salicaceae: *Salix* spp.).

Habitat & Flight Season: along alpine ridges and dwarf willows; flies mostly in July.

Western Meadow Fritillary
Boloria epithore
Wingspan: about 40 mm

The Western Meadow Fritillary does indeed frequent meadows, but not the sorts of aspen meadows that Meadow Fritillaries seem to prefer. For this reason, they are not really eastern and western equivalents of one another. The Western Meadow Fritillary occurs as two subspecies in British Columbia. The northwestern subspecies, which is smaller, darker and rare, is named *B. e. sigridae*, after the wife of lepidopterist Jon Shepard. The southern subspecies is *B. e. chermocki*. This species is most closely related to the Meadow, Frigga and Dingy Fritillaries, and is thus part of the group I call "bubbly" in reference to the smooth contours of their underwing markings.

Similar Species

Meadow Fritillary

Also Called: Pacific Fritillary; *Clossiana epithore*.

ID: a bubbly patterned, purplish hind underwing, and a rounded, not angular, front wing tip distinguishes this species; sexes are similar.

Similar Species: *Meadow Fritillary* (p. 214): has an angular front wing tip.

Caterpillar Food Plants: violets (Violaceae: *Viola* spp.).

Habitat & Flight Season: meadows, usually coniferous forests; adults fly in May at low elevations, in August and early September at high elevations, and in between, in between.

Polar Fritillary

Boloria polaris

Wingspan: about 40–45 mm

Polar Fritillaries share with Arctic Fritillaries the distinction of being our most cold-hardy butterflies. Both extend farther into the Canadian Arctic than any other species. In British Columbia, they live only on the coldest, highest peaks, in places where you might not ever think to look for butterflies at all. It takes the caterpillars two years to reach adulthood, and in this part of the world, the adults are found in odd-numbered years only. Presumably, the even-year cohort, if it ever existed, was wiped out by a particularly harsh season, and has not yet reestablished itself through caterpillars that develop in one or three years. In the high Arctic, it is not unknown for this species to take three years to reach adulthood. Like many of our most northern butterflies, this species is also found in the Old World, in the arctic regions of northern Norway and Siberia.

Also Called: Polaris Fritillary; *Clossiana polaris*.

ID: the hind wing underside is generally brownish, with an angular pattern, and a row of dark, post-median spots bordered by white; sexes are similar.

Similar Species: none of the other angular-patterned fritillaries have the row of post-median spots adjacent to white.

Caterpillar Food Plants: dwarf willows (Salicaceae: *Salix* spp.), or possibly dryas (Rosaceae: *Dryas* spp.).

Habitat & Flight Season: found at the tops of high mountains in B.C.; flies from mid-June to late July.

Alberta Fritillary
Boloria alberta
Wingspan: about 45 mm

What could be more obvious than the fact that a butterfly named *Alberta*, originally discovered and described from Alberta (in 1890, when Alberta was still a provisional district and not yet a province), is in fact named after Alberta? With all due respect to lepidopterists Crispin Guppy and Jon Shepard, their argument (in *Butterflies of British Columbia*) that this species is named for Prince Albert (or for one of Princess Louise's names, as was the province)

Similar Species

Dingy Fritillary

Beringian Fritillary

is not at all convincing to me. I am especially concerned that the use of "Albert's Fritillary" carries with it the risk of annoying butterfly-loving comrades to the east. Both British Columbia and Alberta can be proud of the number of species of butterflies and moths named in their honour, and there is little point in quibbling over a name such as this. At the very least, one would expect that taxonomist W.H. Edwards, who named the species, would certainly have realized that the name could have a double meaning. That aside, what the Alberta Fritillary lacks in colour it makes up for in esteem. Found high in the Rockies, along with the Astarte Fritillary, it is one of the "great northern rarities" about which western Canadians can boast to their butterfly buddies elsewhere.

Also Called: Albert's Fritillary; *Clossiana alberta*.

ID: an extremely sparsely marked, washed-out looking fritillary; sexes are similar.

Similar Species: *Dingy Fritillary* (p. 218): and *Beringian Fritillary* (p. 225): both similarly washed out, but neither is similar in any detail.

Caterpillar Food Plants: probably dryas (Rosaceae: *Dryas* spp.).

Habitat & Flight Season: alpine tundra, especially along scree slopes; flies almost always in July.

Freija Fritillary
Boloria freija
Wingspan: about 40 mm

The Freija Fritillary is the first fritillary of spring, and that alone gives it a special place in the hearts of butterfly fans. I always find it surprising to encounter a lesser fritillary so early in the season (even though I should know better), especially alongside such early-season butterflies as Spring Azures. The Freija Fritillary is named for the Norse god of love and fertility, and her name sometimes appears as "Freya," or "Freiya." It is interesting that the British Columbia populations of this species are not at all different from those in Scandinavia, and do not warrant a separate subspecies name. Perhaps this is a more recent immigrant to North America than some of the other fritillaries, and hasn't had time to diverge from its Old World cousins.

Also Called: *Clossiana freija*.

ID: the hind wing underside has a long white triangle in about the middle of the angular wing pattern; sexes are similar.

Similar Species: *Beringian Fritillary* (p. 225): similar, but larger and darker.

Caterpillar Food Plants: dwarf blueberry (Ericaceae: *Vaccinium caespitosum*), and possibly other members of this family.

Habitat & Flight Season: another boreal forest species, found in meadows and open places, usually flies in late May and early June, but sometimes as early as April, or as late as September.

Beringian Fritillary
Boloria natazhati
Wingspan: about 45–50 mm

This is a rare, northern butterfly that exists in widely separated populations in the Yukon, the western Northwest Territories and northern British Columbia. There are undoubtedly more such populations waiting to be discovered. Interestingly, Beringian Fritillaries are very closely related to Freija Fritillaries, and it is thought (by lepidopterist Jon Shepard and two of his colleagues) that they evolved very recently from a Freija-type ancestor. How interesting that both the typical Old World Freija and its dark, enormous cousin should occur in the same general area today. Reconstruction of what are known as historical biogeographic patterns is one of the main ways in which the otherwise dry process of naming and distinguishing butterflies takes on an interesting complexity. This bit of interesting biology is much easier to overlook when one lumps the Beringian Fritillary in as a subspecies of the Freija, and this is a good example of how names can affect the way we see living things. When there is interesting information to be preserved, I'm all in favour of "splitting."

Also Called: Cryptic Fritillary.

ID: a large, dark, lesser fritillary with an angular underwing and a central white triangle on the hind wing underside; sexes are similar.

Similar Species: *Freija Fritillary* (p. 224): smaller and lighter in colour.

Caterpillar Food Plants: known to feed on dryas (Rosaceae: *Dryas* spp.).

Habitat & Flight Season: alpine areas, above the treeline; flies in June, July or early August, depending on the weather.

Astarte Fritillary
Boloria astarte
Wingspan: about 45–50 mm

The Astarte Fritillary is found in both the Coast Ranges and in the Rocky Mountains, in British Columbia. It appears to take two years to reach adulthood, and in the Coast Ranges the adults are only seen in even-numbered years. In the Rockies, it often occurs alongside the equally tough Alberta Fritillary, although the Astarte seems to be

Similar Species

Dingy Fritillary

Beringian Fritillary

able to survive at slightly lower elevations. The name "Boeber's Fritillary" is used by those who believe that this butterfly and the Distinct Fritillary are both subspecies of a single species, one which also occurs in the Old World. They are considered to be identical in all features except their wing pattern (the Distinct is more dusky and more subdued in its colours), and also identical to their close relatives in Siberia, which also go by the name Astarte. This may well become the standard view, but for the moment I prefer to recognize them as separate species, in a bold attempt to predict the future of lepidopterist thought. In all respects, including size, colour and difficulty of encountering, this is our finest lesser fritillary. It's not one that gets a lot of attention, but the Astarte Fritillary is surely a butterfly that aficionados soon recognize as one of the most sought-after species in the province.

Also Called: Boeber's Fritillary; *Clossiana tritonia*.

ID: a relatively large lesser fritillary with a wavy, white band through the hind wing underside; sexes are similar.

Similar Species: *Dingy* (p. 218) and *Beringian Fritillaries* (p. 225): the Astarte has a distinct underside and is noticeably larger than its relatives.

Caterpillar Food Plants: known to feed on matted saxifrage (Saxifragaceae: *Saxifraga bronchialis*).

Habitat & Flight Season: alpine areas, above the treeline; flies in July, and early August.

Distinct Fritillary

Boloria distincta

Wingspan: about 45–50 mm

Yet another of the "great northern rarities," the Distinct Fritillary is a butterfly I first learned of through my friends Felix Sperling and Gerry Hilchie. In the late 1970s, they made a trip to the Yukon in search of butterflies, and were delighted when they caught some Distinct Fritillaries on top of a remote mountain. Later, in Whitehorse, they stopped in at a government office to look up the name of the mountain, only

Similar Species

Freija Fritillary

Beringian Fritillary

to discover that it had not been named. A clerk handed them a form and suggested that they name it themselves. So, these two lepidopterists coined the name "Distincta Peak," which still applies to this mountain, situated well up the Dempster Highway in the Ogilvie Mountains of the northwestern Yukon. The name *distincta* refers to the distinctness of this butterfly—something that not all lepidopterists agree upon. See the Astarte Fritillary account on p. 226 for an explanation of why these two butterflies could be considered part of a single species.

Also Called: Distinct Astarte Fritillary or Boeber's Fritillary; *Boloria astarte distincta*.

ID: a large, dark lesser fritillary with an angular but somewhat washed-out looking underwing, and a small, central, white triangle on the hind wing underside; sexes are similar.

Similar Species: *Freija Fritillary* (p. 224) and *Beringian Fritillary* (p. 225): generally similar, but not in detail.

Caterpillar Food Plants: may feed on dryas (Rosaceae: *Dryas* spp.).

Habitat & Flight Season: in the alpine at high elevations; adults are found mainly in July.

Titania Fritillary

Boloria titania

Wingspan: about 40–45 mm

This is a familiar lesser fritillary to most butterfly fans because it is both widespread and at home below treeline. I associate it with mid-summer outings, and the peak of the mountain butterfly season. For a while, when we were using "Arctic Fritillary," I really missed the name "Titania Fritillary," although I realize that it had been dropped for good reason. In the 1998 book, *Systematics of Western North American Butterflies*, Jon Shepard presented arguments that the real Titania Fritillary lives only in the Old World. More recent work, however, by Thomas Simonsen, suggests that for the moment, at least, we should retain the Titania name, awaiting further study because it seems to him that both Titania and the Arctic Fritillary (*B. chariclea*) exist in the Old World. "Titania" refers to the Titans of Greek mythology, but I just like the way it sounds.

Similar Species

Freija Fritillary

Also Called: Arctic Fritillary or Purplish Lesser Fritillary; *Clossiana chariclea* or *Clossiana titania*.

ID: hind wing underside is purplish with a unique, angular pattern; sexes are similar.

Similar Species: *Freija Fritillary* (p. 224): larger and more prominent white triangle in the centre of the hind wing underside.

Caterpillar Food Plants: probably willows (Salicaceae: *Salix* spp.).

Habitat & Flight Season: mostly moist meadows; generally flies from late July to early September.

Crescents and Checkerspots
(Subfamily Melitaeinae)

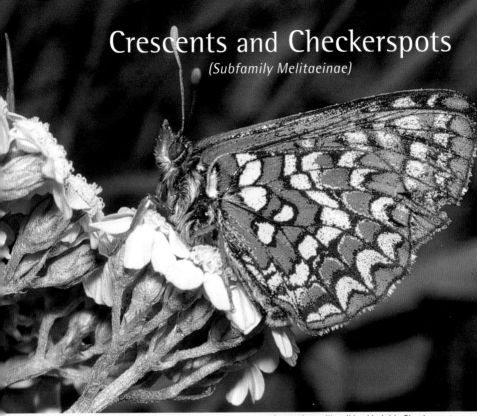

A worn but still striking Variable Checkerspot

In general, these butterflies are lesser fritillary–sized or smaller, and like the fritillaries, they are usually orange and brown. The wing patterns of crescents and checkerspots, however, are not particularly fritillary-like; many are distinctively checkered or lacey in appearance. As well, none of the checkerspots or crescents have silver spots on their underwings. The crescents (genus *Phyciodes*) get their name from a single, somewhat semi-circular spot along the outside margin of the hind wing, and are generally smaller than the checkerspots. The checkerspots have high-contrast underwings, typically marked by bands of white spots alternating with bands of orange spots, with all spots outlined in black. They are pretty! Within the checkerspots, we have two genera: *Chlosyne*, which have much the same overall shape as the crescents, and *Euphydryas*, which have more elongated wings overall.

Male Northern Crescent

Northern Crescent

Phyciodes cocyta

Wingspan: about 35–40 mm

Fresh, male Northern Crescents are some of my favourite butterflies. They look so very elegant, with their wings spread to bask on a flower or low shrub. The females are lovely, too, but there is something about the colour pattern of the male that is particularly appealing.

Crescents are difficult butterflies taxonomically. Not only is it tough to identify each and every member of this species, there is also immense confusion in the literature as to whether this butterfly and the Pearl Crescent (*P. tharos*) are two separate species, or one. Fortunately, the

Similar Species

Field Crescent — Tawny Crescent — Pale Crescent — Mylitta Crescent

Northern Crescent

British Columbia representatives of this "complex" appear to form one species, with the so-called true Pearl Crescent occurring on the Alberta side of the Rockies, south of the range of the Northern Crescent. It is, however, sometimes difficult to tell male from female Northern Crescents, as some males have quite a bit of orange in their outer wing margins. Compare the illustration on page 232 and the photo on page 231 to see what I mean.

Also Called: Pearl Crescent, Northern Pearl Crescent, Pasco Crescent, Orange Crescent or Pearly Crescentspot; *P. pascoensis*, *P. selenis* or *P. morpheus*.

ID: a small orange and black butterfly with a somewhat lacy pattern near the base of the wings; club of the antenna is orange on the tip and on the underside. *Male:* slightly smaller, with less contrast between light and dark orange on the upperwing; crescent spot often obscured by the thumbprint. *Female:* slightly larger, with both pale and darker orange areas on the upperwing; white crescent spot in the thumbprint.

Similar Species: *Field Crescent* (p. 236) and *Tawny Crescent* (p. 234): both a bit darker on the topside, the Field more so than the Tawny. *Pale Crescent* (p. 238) and *Mylitta Crescent* (p. 240): both paler orange, with a more checkerspot-like wing pattern.

Caterpillar Food Plants: asters (Asteraceae: *Aster* spp.).

Habitat & Flight Season: meadows, fields and poplar woods; they appear in one brood in northern regions of the province in June and July, and as two broods in southern regions, the second in the early fall.

Tawny Crescent

Phyciodes batesii

Wingspan: about 32–40 mm

In British Columbia, this butterfly is found only in the Peace River and the Liard River areas. Thus, it is a species of concern from a conservation point of view, but at least it isn't common and widespread enough to cause much identification confusion with the other

Similar Species

Field Crescent

Northern Crescent

crescents! Watch for it in other places, however, just because it is such a difficult species to identify.

The scientific name of the Tawny Crescent honours Henry Walter Bates, a great figure in the study of butterflies, and nature in general. Bates worked extensively in the tropics, and is the namesake of Batesian mimicry, in which a distasteful organism mimics a toxic one. The Tawny Crescent is not involved in any such mimicry, although many tropical crescents do indeed participate in such mimicry "rings."

ID: a typical crescent, like the Northern, but with slightly more extensive dark markings on the topside; tawny-coloured crescent spot; club of the antenna is black and white, without orange markings; male is slightly smaller than the female.

Similar Species: *Field Crescent* (p. 236): nearly impossible to separate from a Tawny. (Check the range maps to see if they occur together in your area.) *Northern Crescent* (p. 232): has either a white crescent spot (in females), or a spot that is obscured by the grey-purple thumbprint (in males).

Caterpillar Food Plants: asters (Asteraceae: *Aster* spp.).

Habitat & Flight Season: aspen poplar forests and meadows; flies from late June through July.

Field Crescent

Phyciodes pulchella

Wingspan: about 35–40 mm

The Field Crescent is probably the second most abundant crescent species in British Columbia, next to the Northern Crescent. It is a familiar butterfly of mountain meadows. Elsewhere, in its extensive range, it is found in a vast array of habitats, and everyone who claims to know this species in British Columbia should be prepared for some surprises when they travel. The similarity of this species and the Tawny Crescent is not accidental, and they probably evolved from a relatively recent common ancestor. Some lepidopterists believe that the two species can, and do, interbreed in some areas as

Similar Species

Tawny Crescent

Northern Crescent

well. To be honest, if you identify one crescent with the help of one book, it is pretty easy to identify it to the species level. If you use several books to identify several crescents, things start getting mighty fuzzy. Should we hold out hope for sophisticated DNA analysis, and how it might some day be possible to catch a crescent, pull off a bit of wing, put it in a handheld "DNA bar-code reader" and get an answer on a backlit screen? Well, taxonomist Niklas Wahlberg in Sweden has bad news—even with the best DNA techniques available today, many of the crescents he has examined wind up being misidentified (or at least identified as something other than what their wing patterns tell us). The solution? Take a deep breath, write "unidentified crescent" in your notes, and live in harmony with the complexity of nature.

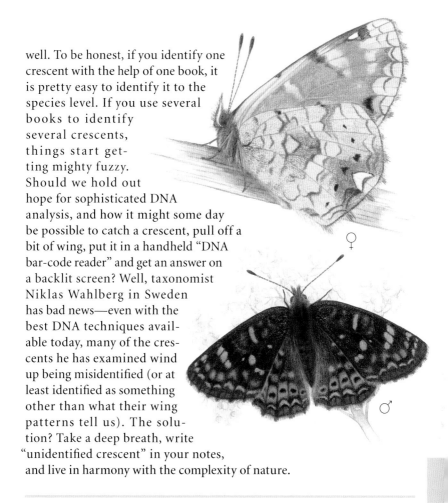

Also Called: Field Crescentspot; *Phyciodes campestris*, *Phyciodes pratensis*, *Phyciodes pulchellus* or *Phyciodes montana*.

ID: a darkish crescent; no orange on antenna club; male is slightly smaller than the female.

Similar Species: *Tawny Crescent* (p. 234): difficult to distinguish the two; very similar, especially on the underside. *Northern Crescent* (p. 232): light crescent spot, and a distinctly different upperwing pattern.

Caterpillar Food Plants: asters (Asteraceae: *Aster* spp.).

Habitat & Flight Season: fields, meadows and roadsides; generally flies in June and July, although populations in warmer places can emerge in May, while those in cooler places may wait until August.

Pale Crescent
Phyciodes pallida
Wingspan: about 45 mm

Lepidopterist Robert M. Pyle has commented that the short flight season and localized nature of this butterfly make it a tough one to find for most naturalists. Because they spend the winter as fourth instar larvae (and remember, the caterpillars have to go through five instars before they form a pupa), they are ready to emerge as adults quite early in the season in some areas. Like other crescents, they are also easy to misidentify. Watch out for mistaking large Mylitta Crescents, or even particularly pale Tawny or Field Crescents, for this butterfly. I have photographs of what looks to be a perfectly normal Pale Crescent, quite a few hundred kilometres to the east of this species'

Similar Species

Mylitta Crescent

Pale Crescent

known range (at Medicine Hat, Alberta), found among quite a few normal looking Tawnies. For that reason, I'm pretty sure it, too, is a Tawny. The same day I found the so-called Pale Crescent, I also found another that was almost entirely black. Fascinating. In Victorian England, butterfly enthusiasts would purposely change the temperature at critical times during a butterfly's pupation, in the hopes that they might get an odd-looking specimen as a result. We should always remember that nature might try the same tricks on us. The resemblance between the Pale and Mylitta Crescents is, however, more straightforward and comes from the fact that they are closely related.

Also Called: Pale Crescentspot; *Phyciodes pallidus* or *Phyciodes barnesi*.

ID: a medium-sized, mostly orange, more or less checkered butterfly; the lightest and biggest of our crescents; male has slightly darker ground colour, with fewer black markings than the female.

Similar Species: *Mylitta Crescent* (p. 240): very similar, but noticeably smaller.

Caterpillar Food Plants: Wavyleaf Thistle (Asteraceae: *Cirsium undulatum*).

Habitat & Flight Season: open, hot, dry areas and streamsides; flies over a relatively short period, anywhere from May to July.

Mylitta Crescent
Phyciodes mylitta
Wingspan: about 35 mm

Like the Pale Crescent in the previous description (and all butterflies, for that matter), the Mylitta is probably also influenced by temperatures during its pupal period. It has been suggested that cool weather results in darker butterflies with lighter-coloured spots, so take care when identifying them. Mylitta Crescents are found in an amazing variety of habitats, and in the warmer southern parts

Similar Species

Northern Crescent   Tawny Crescent   Field Crescent   Pale Crescent

of the province you should have plenty of opportunity to encounter this species. Watch for Mylitta Crescents on a variety of flowers (not just thistles), and follow the cruising males until they stop to bask. Although crescents in general are prone to open-winged basking (and are therefore easy to photograph), the males are patrolling butterflies, not perching butterflies. I like to point out that the crescents are among the smallest gliding animals that have ever flown. Watch as they fly—a burst of wingbeats is followed by a short but very precisely controlled glide. Try to make a decent paper airplane the size of a Mylitta Crescent, and you'll realize how marvelous this really is. Most gliding animals are distinctly on the large end of the flying spectrum, and the largest fliers that ever lived (such as the extinct flying reptile *Quetzalcoatlus*) were probably gliders, not flappers (let alone flutterers!). The Mylitta Crescent has likely inhabited British Columbia for only the last 100 years, following the spread of the Canada thistle.

Also Called: Thistle Crescent.

ID: somewhat smaller and lighter than the rest of the crescents; male has fewer dark markings than the female.

Similar Species: lighter than the *Northern*, *Tawny* and *Field Crescents* (pp. 232–37), and smaller than the *Pale Crescent* (p. 238), although most similar to the latter.

Caterpillar Food Plants: Canada Thistle (Asteraceae: *Cirsium arvense*).

Habitat & Flight Season: generally found in dry, open places, flies in both April and May, and June and July (two broods).

Northern Checkerspot

Chlosyne palla

Wingspan: about 35–40 mm

♂
♀

This butterfly, like the crescents, is part of a confusing "complex" of similar and closely related species. In fact, each of the three genera in the subfamily Melitaeinae in British Columbia contains at least one such "mess," making this our most difficult group of butterflies to identify, by far. The Northern, Rockslide and Hoffmann's Checkerspots not only look alike, they may also interbreed to produce hybrids. And when someone like me advises that it's OK to put "unidentified *Chlosyne*" in your field notes, not everyone feels better.

Setting aside the taxonomic confusion in this group, there are still a few things we can say about their

Similar Species

Rockslide Checkerspot

Hoffman's Checkerspots

lifestyles. First, they seem to be especially attracted to yellow flowers in the aster and daisy family (referred to as "composites" in older books). Their floral preferences make for especially appealing backgrounds in photographs. They also bask with their wings spread, further adding to their appeal. After all, a butterfly that is difficult to identify doesn't have to be dull-coloured. These are beauties, pure and simple. And finally, they like mud-puddles, making them easy to spot as you drive slowly along a gravel or dirt road when the sun emerges after a good rain.

Also Called: Pale Checkerspot or Creamy Checkerspot; *Melitaea palla* or *Charidryas palla*.

ID: a medium-sized, darkish butterfly with both light and dark orange "checkerspots" on the upperwings; sexes are similar, although some females are marked with white, not orange, spotting.

Similar Species: very similar to the other members of its genus in B.C., both of which are found at higher elevations.

Caterpillar Food Plants: asters and rabbitbrush (Asteraceae: *Aster* spp., *Chrysothamnus* spp.).

Habitat & Flight Season: in open meadows and along mountain streams, below 1250 m elevations; flies mostly in June and July.

Rockslide Checkerspot

Chlosyne whitneyi

Wingspan: about 35–40 mm

Not many people have seen this butterfly. After all, it is one thing to go looking for butterflies above treeline, spending your time on rockslides rather than hiking around in nice, flat, mountaintop plateaus. The Rockslide Checkerspot somehow manages to survive in this habitat, as long as the food plants of its caterpillars are able to gain a root-hold, and as long as there are occasional flowers poking up among the rocks to provide the adults with nectar. It may well be that the uneven surface of a rockslide provides the adults and caterpillars with warm microhabitats, sheltered from the alpine winds, in which they can bask and feed in relative comfort. In any event, the Rockslide Checkerspot is the only truly alpine member of its genus, the others preferring to live below treeline. I also like the way lepidopterist Robert M. Pyle has characterized this butterfly, when he writes that it "looks worn when it is fresh." Some butterflies never quite capture the heights of beauty that we seem to expect of them!

Also Called: Whitney's Checkerspot, Damoetus Checkerspot or Alpine Checkerspot; *Charidryas damoetus* or *Charidryas whitneyi*.

ID: very similar to the Northern Checkerspot, but with more extensive white on the underwing surfaces; sexes are similar.

Similar Species: more white on the underside than the other members of its genus in B.C.; lives at higher altitudes than the *Northern Checkerspot* (see p. 236), and farther north than the *Hoffmann's Checkerspot* (see p. 239).

Caterpillar Food Plants: unknown in B.C.; alpine fleabane and alpine goldenrod in the United States (Asteraceae: *Erigeron alpiniformis* and *Solidago multiradiata*).

Habitat & Flight Season: found on rockslides above the timberline; flies from mid-July to early August and into September in some places.

Hoffmann's Checkerspot

Chlosyne hoffmanni

Wingspan: about 35–40 mm

This butterfly is best distinguished by its habitat. The entire geographic range of the Hoffmann's Checkerspot is remarkably small, and takes the form of a north-south band through the Cascade and Sierra Nevada mountains from British Columbia to California. In British Columbia, it is known only from a small area of the Cascades, just north of the U.S. border, in Manning Provincial Park. It is considered a species of special concern here, but mainly because of its small range, not because of any particular threat to the butterfly.

Hoffmann's Checkerspots hibernate as third instar caterpillars and, therefore, have some growing to do in the spring before they become fourth and fifth instar caterpillars, and then pupae and adults. This probably helps time the emergence of the adults to coincide with warm weather and the presence of plenty of nectar-bearing flowers.

Similar Species

Northern Checkerspot

Also Called: Aster Checkerspot or Pacific Checkerspot; *Charidryas hoffmani*.

ID: another typical *Chlosyne* checkerspot; sexes are similar.

Similar Species: similar to the *Northern Checkerspot* (p. 242) and *Rockslide Checkerspot* (p. 244), but most like the *Northern Checkerspot*. It is often the case in Hoffmann's, however, that the pale mid-wing row of spots on the upper forewing merges with the darker orange row of spots next to it, a condition that is rare among Northern Checkerspots.

Caterpillar Food Plants: asters (Asteraceae: *Aster* spp.).

Habitat & Flight Season: flies from late June through July, in meadows and valleys between 1250 and 1900 m in elevation.

Gillett's Checkerspot

Euphydryas gillettii

Wingspan: about 35–45 mm

In British Columbia, this butterfly is found only in the extreme southeast corner, which is a terrible shame. Its range is quite restricted overall, and this is a butterfly of the southern Rockies in Canada, and the northern Rockies in the United States. I personally find it interesting that more butterflies do not share this range pattern because so many of our species are mountain specialists. In *Butterflies*

Similar Species

Variable Checkerspot

Edith's Checkerspots

of Alberta (1993), I suggested, as a mnemonic for remembering field marks, that the Gillett's Checkerspot might be called the "Orange Admiral" because of its wing pattern. Real admirals (and admirables, for that matter) have much simpler patterns, and none of the checkered appearance of the Gillett's. No, the truth is that the Gillett's Checkerspot is indeed a checkerspot, and more importantly, it has relatives in the Old World, quite distinct even from its confusing genus-mates here in North America. For this reason, some prefer to place it in a separate genus (*Hypodryas*) from the Variable Checkerspot and the Edith's Checkerspot. I should also mention that many authors have commented on the relatively slow, weak flight of this butterfly.

Also Called: Gillette's Checkerspot or Yellowstone Checkerspot; *Hypodryas gillettii*.

ID: is the most easily identified checkerspot because of the orange "admiral band" that runs through both wings; male has a more pointed forewing than the female.

Similar Species: other members of the *Euphydryas* genus are similar, but not easily confused. The upperside of the Gillett's is distinctive, but the underside is quite similar to the *Variable* (p. 250) and the *Edith's* (p. 248).

Caterpillar Food Plants: bracted honeysuckle (Caprifoliaceae: *Lonicera involucrata*).

Habitat & Flight Season: moist, open areas in the mountains, usually near flowing water; flies from June to early August.

Edith's Checkerspot

Euphydryas editha

Wingspan: about 30–45 mm

To understand this butterfly in British Columbia, it is helpful to think of it in terms of two different subspecies. The first, "Taylor's Checkerspot" (*Euphydryas editha taylori*) is the low-elevation, coastal subspecies, now restricted to Hornby Island, and an endangered butterfly. Elsewhere, global warming may have contributed to this butterfly's decline, but in British Columbia lepidopterists Crispin Guppy and Jon Shepard argue that habitat destruction and the introduced plant scotch broom are the culprits.

The high elevation subspecies is "Bean's Checkerspot" (*Euphydryas editha beani*). I place the English name in quotation marks because I believe strongly that English names should be given to species alone, not subspecies. Perhaps a better alternative might be "the *beani* race of the Edith's Checkerspot." Whatever you choose to call it, this butterfly is doing well in its sub-alpine meadow and alpine habitats.

Edith's Checkerspot has been the focus of a great deal of study by the

Similar Species

Variable Checkerspot

famous lepidopterist, conservationist and author Paul Ehrlich, and his students and colleagues. They have found many interesting things about the Edith's Checkerspot, including the fact that many caterpillars need to shift from one host plant species to another when the first becomes too dry to eat. They have also seen populations fluctuate to extinction, only to be recolonized later. Increasingly, however, the number of local extinctions is greater than the number of recolonizations, and it appears that climatic warming may be threatening this species. Most famously, Ehrlich and his crew tried unsuccessfully to capture all the butterflies in a small population so that they could study its reestablishment. Their failure to make an impact on even this small population led to the widespread notion that butterfly collectors are not likely to be much of a threat to butterfly populations. The reason for this may be the fecundity of a female Edith's Checkerspot, laying up to 1200 eggs in a week-long lifetime.

Also Called: Ridge Checkerspot.

ID: a checker-spotted butterfly with more orange and white on the wings than black; has an "Editha line"—a black line that runs through the mid-wing row of red spots on the underside of the hind wing; male forewing tip is slightly more pointed than the female's.

Similar Species: *Variable Checkerspot* (p. 250): most similar checkerspot; most coastal *Variable Checkerspots* are darker than Edith's; most *"Anicia" Variable Checkerspots* are larger; Edith's usually appears in flight a few weeks earlier than the Variable where they both occur.

Caterpillar Food Plants: paintbrush, beardtongue, lousewort (Scrophulariaceae: *Pedicularis spp.*), plantain (Plantaginaceae: *Plantago* spp.).

Habitat & Flight Season: flies in April and May in meadows on Hornby Island, and flies in July and August in open areas, high in mountains on the mainland.

Variable Checkerspot

Euphydryas chalcedona

Wingspan: about 40–55 mm

The way I see things, this is the "default" checkerspot—if a *Euphydryas* checkerspot isn't clearly a Gillett's or an Edith's, then it probably belongs to this species. It is also important to realize that the species is often split into two. In the Coast Ranges, some authors see only the Variable Checkerspot proper. Lepidopterists Crispin Guppy and Jon Shepard point out that the only reliable means of confirming the identity of these specimens is to dissect the male genitalia—but don't mount it on a microscope slide, or you may distort the vital features! They interpret this procedural difficulty as the reason that other lepidopterists have included the Anicia Checkerspot as a subspecies of the Variable. The Anicia is the eastern

Similar Species

Edith's Checkerspot

of the two component "species" that make up the Variable Checkerspot. In general, the coastal butterflies are more black than orange or white, and butterflies from the Interior or the Rockies are more orange and white than black. Guppy and Shepard appear to be confident that the two species can be distinguished even in the area of overlap, but for the moment, I have followed the majority of lepidopterists in considering all of these butterflies

to comprise a single species. In *Butterflies of Canada*, Ross Layberry, Peter Hall and Don Lafontaine express the exact opposite view, that the genitalia of many individuals in the area of overlap are indeed intermediate in shape. And as a final point of interest, this butterfly is now extirpated from Vancouver Island; however, the former populations on the island were not, apparently, unique.

Also Called: Chalcedon Checkerspot or Anicia Checkerspot; *Euphydryas anicia*, in part.

ID: a perplexingly variable butterfly, even in a single location; in general, these butterflies are more black than orange or white; male forewing is more pointed than the female.

Similar Species: *Edith's Checkerspot* (p. 248): generally lighter than the Variable on the coast, but not elsewhere, and possesses the "Editha line" on the hind wing underside.

Caterpillar Food Plants: known to feed on snowberry (Caprifoliaceae: *Symphoricarpus* spp.), beardtongue (Scrophulariaceae: *Penstemon* spp.), paintbrush (Scrophulariaceae: *Castilleja* spp.), and plantain (Plantaginaceae: *Plantago* spp.).

Habitat & Flight Season: July and August, in a variety of open areas ranging from forest clearings to alpine meadows.

Nymphs
(Subfamily Nymphalinae)

Painted Lady on an aster (above); *a fresh Milbert's Tortoiseshell* (below)

The nymphs are a wonderful group of butterflies. In thinking about how to characterize them, one word comes quickly to mind: exciting. The appeal of the nymphs, to me, lies not just in their beauty and diversity, but also in the fact that they have stout, powerful bodies, and fast, agile flight. Typically, nymphs fly quickly, remaining in one area—their territory—and gliding between bursts of rapid fluttering. They are our most evasive butterflies, to be sure. As a group, the nymphs are a north-temperate subdivision of the brush-foot family, but that does not mean they are all cold-hardy. The ladies, for example, seem not to be able to overwinter here, in any life stage, and re-invade the province each year. When I travel to places like south Texas to participate in butterfly festivals and lead butterfly-

watching groups, I can't help but reflect on the fact that our biggest and showiest nymphs are just as beautiful and impressive as any of the so-called tropical species that people come to see along the Rio Grande.

Anglewings and Their Relatives
(Tribe Nymphalini)

Mourning Cloak on spring leaf litter

Here, I use the term "anglewing" for members of the tribe Nymphalini other than the "ladies" (the commas, tortoiseshells and Mourning Cloak), and the name "comma" only for members of the genus *Polygonia*. The anglewings do indeed have angular, jagged, outer wing margins, which leaves them resembling flakes of bark, at least on their undersides. Anglewings are forest butterflies, for the most part, and they are also long-lived, hibernating as adults. For this reason, their behaviour is a bit more complicated than that of the more typical, short-lived butterflies because they must deal with long-term nutritional needs, the search for nooks and crannies in which to roost and hibernate and live before and after their reproductive season in the spring. Anglewings, as a group, are prone to dramatic annual fluctuations in numbers, and the tortoiseshells are especially well-known for this, as well as the mass migrations that they exhibit in years of extremely high population density. Personally, I find the anglewings fascinating, and whenever someone asks me what my all-time favourite butterflies are, I often name one of the members of this group (not always the same species because there is no rule that says we have to keep the same favourites day after day!).

As for the ladies (genus *Vanessa*), just like ladybugs, half of them are men. This is a handsome assemblage of strongly migratory butterflies. As a consequence of their migratory habits, they are also widespread, and the Painted Lady is found almost everywhere that butterflies can live on earth. Likewise, our Red Admirable is a familiar European butterfly. However, British Columbia's other two species of *Vanessa* are distinctly North American in character. The ladies are relatively easy to identify and fun to watch.

Satyr Comma
Polygonia satyrus
Wingspan: about 40–55 mm

A satyr, in Greek mythology, inhabits forests and is a sort of minor god, with the body of a man, feet like a goat's and a propensity for rowdy and lustful behaviour. Many butterfly and moth names come from Greek mythology, and the word satyr pops up not only in the name of this butterfly, but also in an unrelated subfamily, which we will treat later in the book. But are Satyr Commas rowdy and

Similar Species

Green Comma

Hoary Comma

Oreas Comma

lustful? Sure they are! This is the most common member of its group in British Columbia, and it's easy to get out and find one on your own. Satyr Commas are both territorial and easily spooked, so watch carefully when you see one, and it will probably return and give you many opportunities for a good view. As far as identification goes, this is the easiest of our comma anglewings to recognize. Its only close relative is the Eastern Comma (*Polygonia comma*), the butterfly that gave the group its name. There are no Eastern Commas in British Columbia, but the Satyr Comma does extend through the boreal forest to eastern Canada and the Great Lakes region, where it confuses other people in ways it cannot confuse us. Satyr Commas have one brood per year in most of British Columbia, but along the southern border of the province they manage two generations. In both broods, as well as the single-brooded populations, the adults that emerge in mid- to late summer feed for a while in the fall, then hibernate and mate in the springtime.

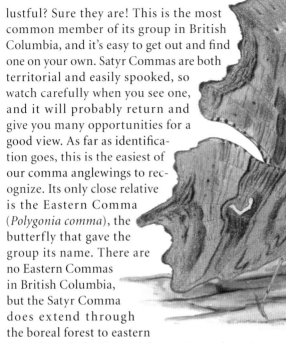

Also Called: Satyr Anglewing, The Satyr, Hope Butterfly or Golden Anglewing.

ID: a relatively light orange, sometimes "golden" anglewing on the topside, with a pale outer border on the hind wing; double dark spots along the middle of the trailing edge of the front wing; wings are warm brown on the underside, not grey, with an exaggerated, somewhat C-shaped, angular "comma" mark on the underside of the hind wing; male is darker on the topside than the female, but both sexes are darker-coloured near the coast and lighter-coloured in the Interior.

Similar Species: generally, other commas are more greyish below, not brown, and are usually darker on the topside.

Caterpillar Food Plants: stinging nettle (Urticaceae: *Urtica dioica*), and hops (Cannabinaceae: *Humulus lupulus*).

Habitat & Flight Season: streamsides and forests; because the adults are long-lived, these butterflies can be found, with luck, throughout the butterfly season.

Green Comma
Polygonia faunus

Wingspan: about 40–50 mm

Back to Greek mythology—a faun was something like a satyr, and Pan (the Greek god of shepherds, pastures and rural life), was a faun. At least that is one take on the name. In other books, I have read that Faunus was the grandson of Saturn and a prophet who visited men in their sleep. This is another very abundant anglewing, and luckily it is also fairly easy to recognize. In the spring, they are found most commonly at sap flows. If you discover a good spot, come back year after year. I have my own favourite sappy places that I visit on the first warm afternoons of spring, to welcome back the butterfly season. By late April, the overwintered adults have mated and the females have laid eggs. From then on, they become harder to find, until the new adults emerge in July.

Also Called: Faun, Faun Anglewing, Faunus Anglewing or Green Anglewing; *P. hylus* or *P. sylvius*.

ID: hind wing topside has a dark central spot, and a row of light spots set in a wide, dark marginal band; this butterfly is said to have more angular wings than other commas, but this is tough to judge in the field; green, lichenlike flecks near the outer margins of the wings on the underside; comma mark in the centre of the hind wing; male has more of the green markings, and a more contrasting underwing pattern overall; female may have very little green, and some are almost plain grey underneath (unlike any other comma).

Similar Species: green flecks or a plain grey underwing make this comma distinctive.

Caterpillar Food Plants: birch and alder (Betulaceae: *Betula* spp. and *Alnus* spp.), and willow (Salicaceae, *Salix* spp.).

Habitat & Flight Season: forests, and along streams and rivers; these butterflies are long-lived and fly throughout the butterfly season.

Hoary Comma
Polygonia gracilis
Wingspan: about 40–50 mm

♂ ♀

Is the Zephyr Comma of the West Coast a separate species from the Hoary Comma in eastern B.C.? Not everyone agrees on this matter, but I am going along with the majority view, that they are one species. If nothing else, this saves us the trouble of distinguishing the two in the northeast part of the province. Increasingly, I am convinced that naturalists

Similar Species

Gray Comma

Green Comma

Oreas Comma

need to accept a simpler taxonomy than the professional system. Among birds and birders, for example, we now know that the Red Crossbill is likely a mix of many similar, related species that are darned-near impossible to tell apart in the field. But the field guides do not show this, and for good reason. Likewise, why agonize over a nearly imperceptible difference between these butterflies, when you could focus instead on other things, such as the often-repeated assertion that this butterfly is more prone to visit flowers than is its relatives. Is this really true? I'm not sure that we know the answer here, and this is something that naturalists could investigate. In Greek mythology, Zephyr was the west wind when the "west" was a heck of a ways to the east of any known records for this species. *Gracilis*, the specific epithet, means slender, in which respect, these butterflies are not in any meaningful sense any different from the other commas.

Also Called: Hoary Anglewing, Zephyr or Zephyr Anglewing; *P. zephyrus*.

ID: to be honest, I cannot recognize this species by its topside pattern; on the underside the wings are grey, and clearly divided into an inner, darker-coloured region and an outer, grey-coloured, or "hoary," region; the comma spot is thin; sexes are similar.

Similar Species: *Gray Comma* (p. 260): most similar; not as distinctly hoary on the outer underwings.

Caterpillar Food Plants: currants and their relatives (Grossulariaceae: *Ribes* spp.).

Habitat & Flight Season: another long-lived, forest-dwelling anglewing, but its range extends higher in altitude than those of its relatives.

Oreas Comma

Polygonia oreas

Wingspan: about 50 mm

Oreads were seductive forest nymphs in Greek mythology, but I find this butterfly more baffling than seductive because I have such trouble recognizing it. Having recommended a relaxed approach to species-level taxonomy in the Hoary Comma, I will continue to do the same here. And again, I want to put forth a bird analogy. Did you know that it is probable that the Canada Goose should really be considered a complex of six related species? If this is true, would you rather try to distinguish them in the field, or admit that some things are beyond the power of mere field naturalists, and that the familiar "Canada Goose" is a landmark for those just learning their birds? To my mind, distinguishing Oreas Commas from Gray Commas is also beyond our powers, although we can admit that the two are different and recognizably so at the extremes. By saying so, I do not wish to pooh-pooh the difference, but merely to say that for my purposes, as a field guide writer, the differences are not worth "teaching" to an audience who will find them more irritating than enlightening.

Similar Species

Gray Comma

Also Called: Oreas Anglewing, Oread Angle-wing, Silenus Anglewing or Dark Gray Anglewing; part of *P. progne*.

ID: on the underwing, look for the light grey V-shaped marks just in from the wing margins; sexes are similar.

Similar Species: *Gray Comma* (p. 260) generally smaller and darker, but can be identical. The Oreas Comma is generally smaller and darker than the Gray; best separated by range.

Caterpillar Food Plants: currants and their relative (Grossulariaceae: *Ribes* spp.).

Habitat & Flight Season: coastal forests, throughout the butterfly year, but usually not abundant.

Gray Comma

Polygonia progne
Wingspan: about 50 mm

♂
♀

L epidopterists John H. and Anna Comstock, the source of many a fine butterfly quotation, referred to the fine white lines that underlie the pattern on the underwing of the Gray Comma as a "woof and warp." They went on to say that "there is nothing from nature's looms that so fills one with the sense of inadequacy of words for description as the under surface of the Gray Comma's wing" (*How to Know the Butterflies*, p. 144). Progne (also spelled Procne) was yet another figure in Greek mythology, a princess who was transformed into a swallow. In fact, you'll find the root "progne" in the scientific

Similar Species

Oreas Comma

names of many swallows, which makes more sense than does the name of this butterfly.

Is this the same species as the Oreas Comma? Under the somewhat smaller and darker Oreas, I argued that we should downplay this distinction. Here, I will present the other side of the coin. It's difficult to argue with Crispin Guppy and Jon Shepard when they say that "The dorsal wing pattern is distinctly different, however, and both species occur together near Quesnel, B.C., in Guppy's front yard"(*Butterflies of British Columbia*, p. 256). Of course, as abstractions, the two are easy to keep straight: the Oreas is a butterfly of the west and the Gray is a more widespread creature of the east. But for those who simply can't tell one from the other (and it's difficult to know how you would, unless you are a collector, and have the luxury of placing a specimen in one series or the other), a species-level identification may never be possible for most individuals.

Also Called: Grey Comma.

ID: the upperwing is bright orange, on the underside, look for grey "Vs"; tan markings at the base of the light-coloured half of the forewing underside; sexes are similar.

Similar Species: *Oreas Comma* (p. 259): smaller and darker, and has a separate geographic range.

Caterpillar Food Plants: currants and their relatives (Grossulariaceae: *Ribes* spp.).

Habitat & Flight Season: forests and along streams; flies from spring through fall, whenever the weather is warm enough.

Compton Tortoiseshell

Nymphalis vaualbum

Wingspan: about 55–75 mm

♂
♀

Similar Species

California Tortoiseshell

This butterfly was named for the town of Compton, Quebec, and in the 1800s, most people also knew what the shell of the Hawksbill Sea Turtle looked like—a lovely, translucent blend of browns, oranges and yellows, often used to make expensive combs. But when it comes to the scientific name, this butterfly, like so many others, is a study in the arcane weirdness that often surrounds butterfly nomenclature. The species was originally described as *N. vau-album*, meaning "white V," and to some this seems reasonable, especially because there is no "J" in either Latin or Greek. But the comma mark, if it is present at all, also looks more like a "J" on one wing, and a backwards "L" on the other. Hence, we also have the names *l-album* and *j-album*, which some think represent species separate from the European species. None of these names are acceptable to the International Code of Zoological Nomenclature, however, because this code does not allow hyphens. Added to this is a much more interesting discussion of whether the Compton Tortoiseshell is more closely related to the other tortoiseshells and the Mourning Cloak, or whether it is more closely related to the commas. Those who believe the latter also use the name *Roddia* for this species. Personally, I think the discussion of this matter in *The Butterflies of Canada* is both reasonable and authoritative, and that is why I use the name *Nymphalis vaualbum* instead.

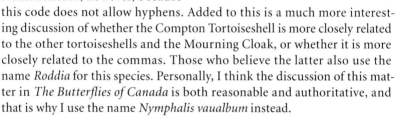

This is one of my all-time favourite butterflies. It is big, beautiful, long-lived and interesting to watch. Compton Tortoiseshells are also prone to huge bursts in numbers, and mass migrations that can take them hundreds of kilometres (or more) from their birthplace, while in other years they spend their entire lives in the same small patches of forest. Some have suggested that this is our longest-lived butterfly, but in my experience, where it is found alongside the Mourning Cloak, the last of the Cloaks live longer into June than do the Comptons.

Also Called: Compton Tortoise, Compton's Tortoiseshell, Comma Tortoiseshell or False Comma (in Europe); *N. vau-aulbum*, *N. j-album*, *N. l-album* or *Roddia l-album*.

ID: like a huge comma with a white patch on each wing; sexes are similar.

Similar Species: *California Tortoishell* (p. 264): the most similar; smaller and more uniformly orange.

Caterpillar Food Plants: willow and poplar (Salicaceae: *Salix* spp. and *Populus* spp.), and birch (Betulaceae: *Betula* spp.).

Habitat & Flight Season: a forest butterfly that hibernates as an adult; long-lived and can appear almost any time that butterflies are on the wing.

California Tortoiseshell

Nymphalis californica

Wingspan: about 50–60 mm

♂
♀

This lovely anglewing is a great butterfly to encounter, but its claim to fame comes not from its everyday activities, but from its population fluctuations. In some years, it is almost impossible to find, even in its stronghold in the southern part of the province. In other years, it can, without any forewarning, become immensely common, and sweep across the southern half of the province in a tremendous dispersal flight. During these "outbreaks" (a term which unfairly suggests that something has gone wrong when it happens, which is not the case), the food plants can

Similar Species

Compton Tortoiseshell

become completely defoliated. In the western United States, where these population spikes are even more pronounced, tales of butterfly chaos are told—cars sliding on the bodies of dead butterflies, trees shimmering with pupae, picnickers fleeing from the terror of so many fluttering wings, and so on. But why does it happen? Well, from an entomologist's perspective, this is pretty normal, and it reminds me a lot of the way grasshopper populations behave. Weather fluctuations affect survival of the butterflies and their young, and the condition of the food plants, and when things come together, well, WHOOSH! They come together. And in response to so-called overcrowding, the butterflies wander, in search of new habitat. For an insect, this is normal. For a bird, it would be odd. To my mind, the greatest lesson to be learned from the California Tortoiseshell is when monitoring population levels for conservation purposes, there might not be anything "level" about it at all.

Also Called: Western Tortoise Shell; *Aglais californica*.

ID: an anglewinged butterfly, about the size of a comma, but with a white spot on each wing, no comma spot below and a more uniformly orange upperwing surface; sexes are similar.

Similar Species: *Compton Tortoiseshell* (p. 262): larger, and more heavily marked on the upper surface.

Caterpillar Food Plants: buckbrush (Rhamnaceae: *Ceanothus* spp.).

Habitat & Flight Season: forests, streamsides and forest clearings; occurs throughout the butterfly season.

Mourning Cloak

Nymphalis antiopa

Wingspan: about 50–80 mm

The Mourning Cloak carries with it a few anecdotes that qualify as part of lepidopterist culture. It is a rare and coveted species in Britain, where it is a migrant and not a breeder. There, it is called the Camberwell Beauty, after an early capture at the town of Camberwell. In North America, where it is common, Mourning Cloaks are more closely associated with springtime. Almost every writer on butterflies has waxed poetic on this theme, but in my view none has said it better than lepidopterists John H. and Anna Comstock, in their usual style: "How the winter-tired eyes are gladdened by this courageous flutterer must be known by experience rather than by description" (*How to Know the Butterflies*, p. 149).

There are still some interesting mysteries surrounding this butterfly. The wing pattern, for example, has been interpreted as mimicking a caterpillar (the pale border) on the edge of a leaf, encouraging birds to

Similar Species

White Admiral

peck at the fake caterpillar, and leaving the butterfly unharmed. Personally, I prefer the notion that the bold wing pattern advertises to birds that this butterfly is very fast and agile on the wing, and not worth pursuing. This theory is supported by the existence of a grasshopper, the Road Duster or Carolina Locust (*Dissosteira carolina*), which clearly mimics the Mourning Cloak. The caterpillars of this butterfly live in groups, and are sensitive to sound. I like to think of them as living V-U metres (the needle-based metres on older stereo equipment that indicate sound level). Speak loudly or sing to a group of Mourning Cloak larvae, and their bodies will twitch just like the real thing! Presumably, they use this as a defence against flying parasites.

Also Called: Camberwell Beauty, White Petticoat, Grand Surprise, Antiopa, Spiny Elm Caterpillar or Yellow Edge.

ID: almost unmistakable with maroon wings, large size and a white or yellow outer wing border that make it distinctive; sexes are similar.

Similar Species: *White Admiral* (p. 276): light wing bands are on the middle of the wings, not the outsides.

Caterpillar Food Plants: willows and poplars (Salicaceae: *Salix* spp. and *Populus* spp.), and elms (Ulmaceae: *Ulmus* spp.).

Habitat & Flight Season: not as tightly tied to forests as the preceding species of anglewings, and may be found well into open meadows and fields; flies throughout the butterfly season.

Milbert's Tortoiseshell
Aglais milberti
Wingspan: about 40–55 mm

Like the Mourning Cloak, this species carries with it a certain amount of European baggage. It was named by a man named Godart, for his friend "Mr. Milbert," who was himself a butterfly collector, although otherwise not very famous, despite having gained immortality through this butterfly's name. But many people feel otherwise, and wish to replace the name "Milbert's" with something more descriptive, or contemporary. Immortality is never really what it's cracked up to be. Also, on the European theme, Milbert's Tortoiseshell is most closely related to the Small Tortoiseshell, *A. urticae*. If you are interested, do look for a picture of this butterfly, which is a perfect intermediate between the Milbert's and a typical comma. With the Small Tortoiseshell, and our local species, you can easily see how the commas are related to the tortoiseshells, and how the Milbert's is also reasonably similar to a Mourning Cloak. The evolutionary tree that joins the members of this group just about leaps off its wings, at least to the trained eye. Milbert's Tortoiseshells have up to three broods per year in eastern North America, but in British Columbia they share the same life history as all of their close relatives in the anglewing group.

Also Called: Nettle Tortoise Shell, Fire-rim Tortoise Shell or American Tortoiseshell; *Nymphalis milberti*.

ID: another unmistakable species once you get to know it; the "fire rim" and blue spots on the topside, and the very dark underside, are both easy to recognize in the field; sexes are similar.

Similar Species: Not really similar to any other butterfly, but *Mourning Cloaks* (p. 266) are sometimes described as being similar.

Caterpillar Food Plants: stinging nettles (Urticaceae: *Urtica* spp.).

Habitat & Flight Season: in a wide variety of habitats, including the alpine zone; flies throughout the butterfly season.

Painted Lady
Vanessa cardui
Wingspan: about 50–70 mm

For such a familiar butterfly, the Painted Lady has a remarkably tenuous connection to British Columbia. Unlike the anglewings and tortoiseshells, the ladies cannot survive hibernation in such a cold place (although a few Canadian records suggest that it is possible from time to time). Each year, the permanent populations of the southwest deserts of the United States and Mexico collect over the winter

Similar Species

West Coast Lady

American Lady

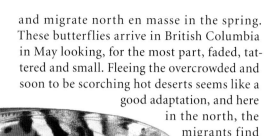

and migrate north en masse in the spring. These butterflies arrive in British Columbia in May looking, for the most part, faded, tattered and small. Fleeing the overcrowded and soon to be scorching hot deserts seems like a good adaptation, and here in the north, the migrants find plenty of food plants for their caterpillars, such that a fresh, beautiful and physically larger summer generation emerges in late June and July. But do they return southward? Nope. In the southwestern United States some do, apparently, but here they just tough it out and eventually expire. They seem to live for the moment, but of course so do we all, whether we know it or not. The name of this butterfly is an odd one, but I like it. And for a cosmopolitan species, it is futile to insist on a single, standardized name. I love to call the odd individual "Cynthia of the Thistle," just to hear the echoes of the 1800s in my own voice. The last time a big wave of Painted Ladies came to Canada, I first noticed them while sitting in an airplane on the tarmac of the Dallas airport. A few months later, right on schedule, their progeny arrived en masse in western Canada.

Also Called: Cosmopolite, Cynthia, Cynthia of the Thistle or Thistle Butterfly; *Cynthia cardui*.

ID: an easily recognized pink-orange and black butterfly with white spots near the tips of the forewings; spot on the forewing margin is white; four eyespots on the hind wing underside; sexes are similar.

Similar Species: *West Coast Lady* (p. 271): spot along the forewing margin is orange, instead of white. *American Lady* (p. 272): has two (not four) large eyespots on the hind wing underside.

Caterpillar Food Plants: thistles (Asteraceae: *Cirsium* spp. and *Carduus* spp.) and other members of the daisy family.

Habitat & Flight Season: in migration, they can be found almost anywhere; migrants arrive in May, and the local generation flies in July and on into the fall.

West Coast Lady
Vanessa annabella

Wingspan: about 40–55 mm

This is another migrant lady, with a life history similar to the Painted Lady, although the migrants come from closer by along the U.S. west coast. And because they come from a nearby source, they seem not to have the same migratory zeal as the Painted Lady, and colonize only the southern parts of British Columbia as a result. This is not to say that long-distance flights never occur in this species; I found one once just west of Edmonton, Alberta (there are records from Saskatchewan as well). The West Coast Lady was formerly thought to belong to the same species as the South American Lady species V. carye, but was separated formally in 1971 by a man named Field, who named the "new" species for his daughter Annabell. She, thus, joined the ranks of Mr. Milbert in butterfly timelessness while also lending a lovely name to a lovely creature. This butterfly is a hilltopper, as are most of its lady relatives, and in mid-summer, it is as common in the alpine zone as the "true" alpine butterflies, if not more so.

Similar Species

Painted Lady

American Lady

Also Called: Western Painted Lady; *V. carye*.

ID: this butterfly is generally orange in ground colour; orange spot along the forewing margin; a noticeably squared-off front wing tip; centre spots on the upper hind wing have a blue centre, corresponding in position to the eyespots on the underside; sexes are similar.

Similar Species: *Painted Lady* (p. 269): larger, pinker and has all white spots near the forewing tip; *American Lady* (p. 272): has two, not four, eyespots on the hind wing underside.

Caterpillar Food Plants: stinging nettle (Urticaceae: *Urtica* spp.) and garden hollyhock (Malvaceae: *Alcea rosea*).

Habitat & Flight Season: found in almost any sunny place, from May to late fall.

American Lady
Vanessa virginiensis

Wingspan: about 40–60 mm

Rare in British Columbia, this is the lady of eastern North America. Like the West Coast Lady, it is a bit more cold-hardy than the Painted Lady, and thus maintains breeding populations farther north in the United States. The west is easily within its reach, but it is not especially to this butterfly's ecological liking. Call me corny, but I like the name "Hunter's Butterfly" as well. Because the ladies are long-lived, this butterfly is one of the most common late fall species in the east during hunting season. Although I am not a hunter myself, I have a vivid mental picture of the camouflage-clad hunter of yesteryear, stopping to puff on his pipe and chew a bit of jerky, then noticing a butterfly perched in a sunny spot beside him, slowly opening and closing its wings. The hunter chuckles that smug sort of chuckle that woodsmen seem to master so easily, self-satisfied that he knows the butterfly's identity. It is the Hunter's Butterfly, and it brings good luck to the hunt. And if you happen to be hunting butterflies, perhaps the deer and the wild turkeys will bring good luck in return.

Also Called: Hunter's Butterfly, American Painted Lady or Virginia Lady.

ID: recognizably a lady, with an orange ground colour on the topside; two (not four) eyespots on the hind wing underside; the large light spot along the forewing margin is usually white, but can be orange; sexes are similar.

Similar Species: the other two ladies are similar, but easily distinguished by the number of eyespots on the hind wing underside.

Caterpillar Food Plants: not known to breed in B.C., but feeds on various daisy family plants (Asteraceae).

Habitat & Flight Season: a rare stray to British Columbia that could turn up anywhere, but more likely in the south.

Red Admirable

Vanessa atalanta

Wingspan: about 50–60 mm

The Red Admirable was so named over 250 years ago in Britain. Another British species, *Ladoga camilla*, was called the White Admiral. Predictably, the two names were easy to confuse, and in North America we now have another species (*Limenitis arthemis*) called the White Admiral, and the Red Admirable has, by most authors, been called the Red Admiral. Those of us who write and teach about butterflies find this annoying, since the two butterflies are not close relatives, and thus the name "admiral" loses its explanatory value. So, lepidopterist Robert M. Pyle has championed the old name, and I have decided to follow his brave lead. This is a lovely butterfly, by any name, and one that overwinters along the West Coast. Unlike the other ladies (and perhaps it should really be called the "Red-banded Lady"), it is not too terribly drawn to hilltops, and males will patrol territories in forest clearings and gardens. Like the other ladies, however, it is indeed migratory. In California, hybrids between Red Admirables and West Coast Ladies have been found, but so far there are no such records for British Columbia.

Also Called: Red Admiral, Nettle Butterfly or Alderman.

ID: a dark brown or black butterfly with a red band through the wings; white spots near the front wing tips; sexes are similar.

Similar Species: this species is distinctive.

Caterpillar Food Plants: nettles (family Urticaceae).

Habitat & Flight Season: a variety of habitats; flies from May on into the fall, with occasional winter sightings.

A lovely, backlit White Admiral

Admirals
(Subfamily Limenitidinae)

These are admirals in the true sense, unlike the Red Admirable in the preceding section. *Limenitis* means "harbour protector" in Latin, and some butterfly authors have suggested that this is what admirals do. Others suggest that the word "admiral" comes from "admirable." In either case, this is a worldwide subfamily of butterflies, of which our province has only two representatives. A third, the Viceroy (which is an orange and black butterfly that shares its mimetic colours with the Monarch) has not been seen in British Columbia since 1930. According to lepidopterists Guppy and Shepard, Viceroys expanded their range to feed on cultivated apples, then disappeared because of pesticides. Viceroys do feed on poplars and willows, however, and with the changes currently happening in climate, pesticide use and butterfly genetics, they may reappear some day. All admirals have caterpillars that look like bird poop when they are young, and they overwinter in a curled-up leaf held together by silk. The adults, although beautiful, are famous for their attraction to rotting fruit, urine and carrion.

Lorquin's Admiral
Limenitis lorquini

Wingspan: about 50–70 mm

Of the two British Columbia admirals, this is the more abundant. It is, in my opinion, one of the signature butterflies of the West Coast as well. Most often, I have encountered it from below, which is to say that these butterflies often perch high in trees, spreading their wings to sun themselves. From these perches, the males are also territorial and often dash out after other butterflies, returning to their favourite perches between battles. With the sun streaming through their wings, the sight of a Lorquin's Admiral can be truly gorgeous. And who was Lorquin? Pierre Joseph Michel Lorquin was a French butterfly enthusiast who lived in California. Even now, if you visit the Los Angeles area, you will find the Lorquin Entomological Society alive and well, and filled with keen lepidopterists. Speaking of California, and places to the south, there is another genus in the admiral subfamily, *Adelpha*, called the "sisters." These great-looking butterflies also possess orange wing tips, and to those of us who travel a lot, the Lorquin's wings are an indication of the close relationship between the two groups.

Similar Species

White Admiral

Also Called: Orangetip Admiral; *Basilarchia lorquini*.

ID: a lovely, large butterfly with a white band through dark wings, and orange wing tips; sexes are similar.

Similar Species: *White Admiral* (p. 276): very similar, but without orange wing tips; hybrids between the two may pop up from time to time.

Caterpillar Food Plants: primarily willows and poplars (Salicaceae: *Salix* spp. and *Populus* spp.), and chokecherry and its relatives (Rosaceae: *Prunus* spp.).

Habitat & Flight Season: in forest clearings and along roads and in gardens; seen primarily in midsummer; a partial second brood extends their flight season into the fall.

White Admiral

Limenitis arthemis

Wingspan: about 50–70 mm

This is another super-good-looking butterfly that perches high in vegetation and defends a territory. Most of the time, it is seen in June and July, but those larvae that have completed a good portion of their development by midsummer will go ahead and pupate, to emerge in late summer or early fall. The others, which comprise the majority, overwinter as partly grown larvae, and emerge the following spring. For such an easy-to-identify, large, attractive butterfly, this one also carries with it a bit of buttergack. In eastern Canada, one subspecies (*L. arthemis astyanax*) has no white band through the wing. Instead, it is a lovely, purplish mimic of the Pipevine Swallowtail, and was long considered a separate species, and called the Red-spotted Purple. Now that we consider it part of the same species as the White Admiral, some have suggested renaming the entire species the "Red-spotted Admiral." Now this sort of change may work with small, obscure species, but for a butterfly as familiar as the White Admiral, it is hopeless to try to change the name. Thus, most of us simply use "White Admiral" for subspecies that have white bands, and "Red-spotted Purple" when we travel to eastern North America.

Similar Species

Lorquin's Admiral

Also Called: Red-spotted Purple or Red-spotted Admiral; *Basilarchia arthemis*.

ID: a large, dark butterfly with a white band through both pairs of wings; numerous red and blue markings near the wing margins; sexes are similar.

Similar Species: *Lorquin's Admiral* (p. 275): has orange wing tips, as do most hybrids between these two species.

Caterpillar Food Plants: willows and poplars (Salicaceae: *Salix* and *Populus* spp.), and birch (Betulaceae: *Betula* spp.).

Habitat & Flight Season: woodland edges and clearings; flies in June and July, some in August and September, from a partial second brood

Satyrs
(Subfamily Satyrinae)

A Ringlet on three-flowered avens

As I explained when describing the Satyr Comma, satyrs, in Greek mythology, were minor woodland deities. This woodsiness, as far as I can tell, is the only reason their name has been associated with butterflies, and to be clear on the matter, the subfamily Satyrinae is not a particularly woodsy group of butterflies. These are generally earth tone–coloured species, marked in browns, greys and dull oranges, often with rounded eyespots near the wing margins. Most satyrs fly in a characteristic bobbing fashion, generally low to the ground. Most of the time, they hold their wings closed above their backs and flap very quickly between "bobs."

For the most part, satyrs are butterflies of open country, and they are most diverse above treeline. Still, a few species are truly forest dwellers, and some are sufficiently rare and elusive to be worthy of great excitement when they are found. The first four species I discuss in the next section are satyrs of low-elevation meadows, and for this reason they are among the satyrs most likely to be encountered by the casual butterfly enthusiast.

Ringlet
Coenonympha tullia
Wingspan: about 30–40 mm

This is a good satyr to start with. It's easy to recognize and does all the typical satyr things: it lives in meadows, has a bobbing flight style and produces caterpillars that eat grasses. To most lepidopterists, there are two species of ringlets in North America—this one and the Hayden's Ringlet (*C. haydeni*) found in Montana, Wyoming and Idaho. Some also recognize a third, salt-marsh species, the Maritime Ringlet (*C. nipisquit*) of New Brunswick and Quebec. Ringlets also occur in Europe, and there the "Large Heath" is considered by most to be the same species as our Common Ringlet. In British Columbia, lepidopterists Crispin Guppy and Jon Shepard have shown that the northern populations of ringlets have genitalia more like the European populations. For this reason, they call individuals found in the north "Northern Ringlets" (*C. tullia*), and the ones in the south "Common Ringlets" (*C. californica*). In my opinion, this arrangement is not yet entirely convincing, and it is also important to remember that some entomologists dislike the use of European names for North American insects, even when they look exactly alike. I don't feel this way myself, so I am calling all the B.C. ringlets *C. tullia*, with no intention of slighting anyone else's entomological conclusions.

Also Called: California Ringlet, Ochre Ringlet, Common Ringlet, Inornate Ringlet or Large Heath; *C. californica*, *C. ochracea*, *C. ampelos* or *C. inornata*.

ID: the upperwing surfaces vary from almost white to plain pumpkin orange; a pale, zig-zagged streak through both wings on the underside; sometimes a small eyespot near the apex of the front wing on the underside; sexes are similar.

Similar Species: none.

Caterpillar Food Plants: grasses (Poaceae).

Habitat & Flight Season: in grassy meadows, at all elevations, mostly in June (except on Vancouver Island where the species, although rare, has a second brood in late summer).

Common Wood Nymph

Cercyonis pegala

Wingspan: about 50–60 mm

This is the wood nymph of eastern North America, where its eyespots are set in a rectangular yellow patch on the wing. Like most eastern butterflies, this makes it easy to identify. In fact, it is a general fact that eastern butterflies are easier to identify than western ones. This is probably a consequence of the more complex topography, and therefore more complex geological history of the West, combined with the fact that the arid nature of the West has resulted in more isolated butterfly populations over the long course of geological time. In any event, the yellow patch on eastern

Similar Species

Great Basin Wood Nymph

Dark Wood Nymph

Common Wood Nymph

members of this species gave it the name "Goggle-eyed Wood Nymph," which in my opinion is the funniest butterfly name of all, if you don't count "Noseburn Wanderer," an out-of-fashion name for a lovely small butterfly found in Texas, among other places. And while I'm on the subject of names, many other authors have insisted on either "woodnymph" or "wood-nymph," but I am convinced that two words clearly communicate the meaning of the name better than a compound noun. Getting back to biology, I should mention that because the Common Wood Nymph is at home in eastern Canada, it predictably favours more moist habitats in western Canada than do its close relatives. In the heat of summer, in relatively lush, grassy places, this is a familiar dark brown butterfly, bobbing along beside such year-round species as the Cabbage White and the Clouded Sulphur.

Also Called: Large Wood Nymph, Ox-eyed Wood Nymph or Goggle-eyed Wood Nymph.

ID: a brown, round-winged butterfly with two obvious eyespots on the forewing, especially on the underside. Of these, the lower (hindmost) eyespot is almost always larger than the upper (foremost) eyespot.

Similar Species: other wood nymphs are smaller, and have lower eyespots that are smaller than their upper eyespots.

Caterpillar Food Plants: grasses (Poaceae).

Habitat & Flight Season: grassy areas, at lower elevations, mostly in July and August.

Great Basin Wood Nymph

Cercyonis sthenele

Wingspan: about 40–45 mm

♂

Unlike the Common Wood Nymph, this is a butterfly of western North America. It was originally described in California, but that population is now extinct. For this reason, some butterfly experts have decided to use the next oldest name, *C. silvestris*, for this species, but there is nothing in biology that says you have to change a species' name when the so-called typical population goes extinct. In a way, this is unfortunate because *silvestris* means "of the forest," and is one heck of a lot easier

Similar Species

Common Wood Nymph

Dark Wood Nymph

Great Basin Wood-Nymph

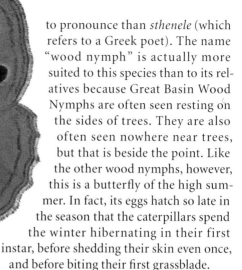

to pronounce than *sthenele* (which refers to a Greek poet). The name "wood nymph" is actually more suited to this species than to its relatives because Great Basin Wood Nymphs are often seen resting on the sides of trees. They are also often seen nowhere near trees, but that is beside the point. Like the other wood nymphs, however, this is a butterfly of the high summer. In fact, its eggs hatch so late in the season that the caterpillars spend the winter hibernating in their first instar, before shedding their skin even once, and before biting their first grassblade.

Because Sthenele was a poet, let me finish with this:

> Sthenele, sthenele,
> I knew it well,
> When S, T and H,
>   Were much easier to spell.

Also Called: Woodland Satyr, Small Wood Nymph or Scrub Wood Nymph; *C. silvestris*.

ID: on the forewing underside, the lower eyespot is the same size or smaller than the upper; the outer half of the hind wing underside is distinctly paler than the inner half.

Similar Species: *Common Wood Nymph* (p. 279): has a larger lower eyespot. *Dark Wood Nymph* (p. 283): has an even smaller lower eyespot, also closer to the wing edge.

Caterpillar Food Plants: grasses (Poaceae).

Habitat & Flight Season: flies mainly in July and August, in sagebrush areas and open coniferous forests.

Dark Wood Nymph
Cercyonis oetus
Wingspan: about 40–45 mm

I wish I could say that there are things to be said about the Dark Wood Nymph that have not been said about the Great Basin Wood Nymph, but apart from the expected taxonomic discussions, most butterfly experts seem to agree that these two butterflies are quite similar in other ways. The Dark Wood Nymph is more widespread east of the Rockies, but that's not something you would notice in British Columbia. It is worth noting, however, that the Dark Wood Nymph is found at higher elevations than

Similar Species

Common Wood Nymph

Great Basin Wood Nymph

Dark Wood Nymph

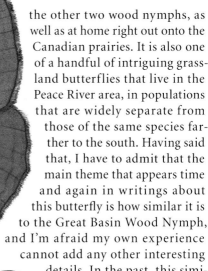

the other two wood nymphs, as well as at home right out onto the Canadian prairies. It is also one of a handful of intriguing grassland butterflies that live in the Peace River area, in populations that are widely separate from those of the same species farther to the south. Having said that, I have to admit that the main theme that appears time and again in writings about this butterfly is how similar it is to the Great Basin Wood Nymph, and I'm afraid my own experience cannot add any other interesting details. In the past, this similarity resulted in both species, in different books at different times, being referred to as the "Small Wood Nymph." Robert M. Pyle, in his typically bold and incisive way, has suggested that W.J. Holland's name, "Dark Wood Nymph" is the better choice for this species. As with "Red Admirable," I am willing to follow Dr. Pyle's lead here, and suffer the inevitable criticisms of my colleagues, many of whom think that "Small Wood Nymph" is now an official name.

Also Called: Small Wood Nymph.

ID: lower eyespot on the forewing below smaller (or absent) and closer to the wing edge than the upper spot. Otherwise very recognizably a wood nymph.

Similar Species: *Common Wood Nymph* (p. 279): larger, with a lower eyespot that is larger than the upper on the topside of wings. *Great Basin Wood Nymph* (p. 281): has a slightly larger lower eyespot that is at the same distance to the wing edge as the upper spot.

Caterpillar Food Plants: grasses (Poaceae).

Habitat & Flight Season: common in sage flats, but also found in fields, prairie and forest meadows; flies in July and August.

Alpines
(Genus *Erebia*)

Common Alpine

Among the deep aficionados of butterflies, the alpines have always had a certain mystique. You might find this surprising because alpines are generally dull in colour, but let's consider the things that make them such favourites. First, there are a couple of species that are common and widespread at low elevations, and, therefore, whet the interest of casual butterfliers. Second, the rest of the genus is found only in the inhospitable reaches of the high alpine, or the arctic tundra, with a few species here and there in places like bog forests. So you have to make a special effort to find them. And when you do find them, some are quite tricky to identify, which is a sure-fire way to make butterfly people feel smart once they get the hang of telling the species apart. To make the alpines even more appealing, there are plenty of species in Europe, where butterfly enthusiasts have an even stronger tradition of Erebiamania, dating back hundreds of years. One of the great classic works in butterfly taxonomy was the *Monograph of the genus* Erebia, written by Brisbane C.S. Warren and published in 1936, which covered all 100 or so species in the group. It was an inspiration to all sorts of butterfly collectors, at a time when butterfly collecting already had a long history. The result of all this was the widespread notion that to be a true butterfly expert, you should really get to know your alpines. Even in today's hands-off butterfly-watching climate, the mystique survives. In fact, the father of butterfly watching in North America, Robert M. Pyle, has long been working on a novel, the working title of which is "Magdalena Mountain," and in which the Magdalena Alpine plays a significant role. Bob is continually rewriting the novel itself, so it has yet to be published. In fact, it is the most famous nonexistent novel I can think of, and almost everyone with an active interest in butterflies has heard of it.

There are three alpine species whose ranges come within spitting (or flitting) distance of the northwest corner of British Columbia at Atlin, near the Yukon border: the Young's Alpine (*Erebia youngi*), the Reddish or Lafontaine's Alpine *(E. kozhantshikovi = E. lafontainei)*, and the Banded Alpine (*E. fasciata*). These are not included in the pages that follow, in part because it is surely the case that anyone keen enough to venture that Far North to look for butterflies would also be keen enough to buy all the other relevant butterfly books as well.

Red-disked Alpine

Common Alpine
Erebia epipsodea
Wingspan: about 35–45 mm

In most books, the Common Alpine is nowhere near the front of the list of alpines, but in my opinion, this is the species you should get to know both first and best because it is common at low elevations, and looks a lot like many of its relatives. If an alpine doesn't look something like the Common Alpine, it probably looks something like the Red-disked Alpine, so get to know these two first, and the rest of the job will become easier. Having characterized the Common Alpine as the archetypical alpine, I should also add that the low-elevation

Similar Species

Vidler's Alpine

Taiga Alpine

Ross's Alpine

Common Alpine

♂
♀

habitat of the Common Alpine makes it a somewhat unusual member of its genus. As lepidopterist Robert M. Pyle says, this may be the only alpine that global warming will treat kindly; warming temperatures may squeeze other alpines off the tops of their preferred, cool mountains. Of course, we have had warm climatic periods in the past as well (about 5000 years ago, especially), and then it is believed that the high-altitude butterflies simply shifted their ranges northward, hopping from mountain peak to mountain peak. Studies of the behaviour of adult Common Alpines have shown that while most stay within 500 m of their birthplace, some disperse as far as 15 km. Thus, they are able to find nearby patches of good habitat, as well as spread their genes between populations and prevent inbreeding in isolated colonies.

Also Called: Butler's Alpine.

ID: a medium-sized brown butterfly with eyespots set in orange patches on all four wings, including the hind wing underside; sexes are similar.

Similar Species: *Vidler's, Taiga and Ross' Alpines* (pp. 289–91): do not have eyespots on the hind wing underside.

Caterpillar Food Plants: grasses (Poaceae).

Habitat & Flight Season: found in open grassy areas; flies from June at lower elevations to July and August in the mountains.

Taiga Alpine
Erebia mancinus
Wingspan: about 35–45 mm

The Taiga Alpine is a bog butterfly, and bog butterflies have a special appeal. It is not at all easy to find them in their heavily treed habitats, and even more difficult to follow one once you do find one of these butterflies. Taiga Alpines make a habit of resting on tree trunks, and they are often found in the same places as Jutta Arctics, another sort of satyr. It is easy to tell the two apart at a distance, however, since the Taiga Alpine shows the bobbing style of flight, while the Jutta Arctic flies much more like a typical brushfoot, with flat-winged glides between flapping bouts. Taiga Alpines also have a habit of dropping to the ground when pursued. Until quite recently, this butterfly was known as the Disa Alpine (*E. disa*). In *Butterflies of Canada*, Ross Layberry and his co-authors decided to distinguish the "true" Disa Alpines of the far northern tundra from the Taiga Alpines of the boreal forest. There are differences between the two species' male genitalia. Personally, I will miss the name Disa. In his classic field guide to eastern butterflies, Alexander Klots even managed to slip a pun past his editors, when he wrote "The subspecies *mancinus* Westwood (TL "Rocky Mountains") of disa cirmcumpolar species has been taken as far southeastward as Smoky Falls, Mattagmai R., Ontario (29 June)" (*A Field Guide to the Butterflies of North America East of the Great Plains* (1951), page 77). For many of us, this is a legendary bit of humour when it comes to disa butterfly.

Similar Species

Ross's Alpine

Also Called: Mancina Alpine; *E. disa mancinus*.

ID: like the Common Alpine, but with smaller eyespots and orange patches, and a small white spot near the centre of the hind wing underside; sexes are similar.

Similar Species: *Common Alpine* (p. 287), have eyespots on the underside of the hind wing. *Vidler's Alpine* (p. 290), and *Ross's Alpine* (p. 291), both lack the white spot on the hind wing underside.

Caterpillar Food Plants: unknown, amazingly enough, but probably grasses (Poaceae) and sedges (Cyperaceae: *Carex* spp.).

Habitat & Flight Season: open peatlands, among black spruce and tamarack trees; flies in late June and July.

Vidler's Alpine
Erebia vidleri
Wingspan: about 35–45 mm

♂ ♀

Like the Common Alpine, you can often find Vidler's Alpine sipping nectar at daisy-family flowers, its wings held open to the sun; however, this is a butterfly endemic to the Pacific Northwest, found only in the Cascades and the Coast Ranges. Oddly, despite the fact that Vidler's Alpine has a unique range, not much has been written in standard butterfly guides about this species. It is a common species in its range, perhaps making it common in collections as well, preventing it from gaining a reputation as a great rarity. In all fairness, *The Butterflies of Canada* calls it the most colourful North American alpine, but let's face it, that's not saying much. Apparently, Vidler was a Russian sea captain, who collected one of the first of these butterflies, which was to become the type specimen. Perhaps such an obscure name has also worked against the reputation of this butterfly, which should be the pride of all naturalists in the Pacific Northwest (perhaps more so, I might add, than the "Oregon Swallowtail"!).

Similar Species

Common Alpine

Taiga Alpine

Ross's Alpine

Also Called: Northwest Alpine or Cascades Alpine.

ID: has checkered wing fringes and no eyespots on the underside of the hind wing; sexes are similar.

Similar Species: *Common Alpine* (p. 287) has an eyespot on the underside of the hind wing. *Taiga Alpine* (p. 289): has a white spot on the hind wing underside. *Ross's Alpine* (p. 291): has very small orange patches around the eyespots.

Caterpillar Food Plants: probably grasses (Poaceae).

Habitat & Flight Season: moist mountain meadows and open subalpine forests; flies in July and August.

Ross's Alpine

Erebia rossii

Wingspan: about 35–45 mm

We know of only a few populations of Ross's Alpine in British Columbia, which are probably isolated islands in the distribution of the species. This butterfly lives in both northern North America and Siberia, and along the Arctic Ocean coast to Baffin Island and the western edge of Hudson Bay. Like the Vidler's Alpine, it is common, but only in the specific areas where you find it, so could it therefore be considered rare, but not really. Like the Common Alpine, it is often found sipping flower nectar, or on mud. The Arctic explorer Captain John Ross led the expedition on which it was first found, but it was named for his nephew, another "Captain Ross," James Clark Ross, instead. The name "Ross" has an arctic association for birders. The Ross's Gull was named for James C. Ross, and to bird watchers, it is one of the most sought-after arctic birds in North America. We should grant the same respect to the Ross's Alpine, regardless of which Captain Ross it was named for (and at this point, I think most people with an interest in such things associate both captains with both animals as well).

Similar Species

Taiga Alpine

Magdalena Alpine

Mt. McKinley Alpine

Also Called: Ross' Alpine or Two-dot Alpine; *E. rossi*.

ID: similar to the Common Alpine, but with smaller orange margins around the very small, upper eyespots; usually only two eyespots; no white spot on the hind underwing; sexes are similar.

Similar Species: *Taiga Alpine* (p. 289): has a white spot on the hind underwing. *Magdalena Alpine* (p. 293) and *Mt. McKinley Alpine* (p. 294): darker but similar, especially in flight.

Caterpillar Food Plants: sedges (Cyperaceae: *Carex* spp.).

Habitat & Flight Season: alpine and subalpine in wet tundra and sedgy shrublands and peatlands; flies in June and July.

Red-disked Alpine
Erebia discoidalis
Wingspan: about 35–45 mm

I n most of Canada, the Red-disked Alpine is a ubiquitous, low-elevation species, found in any moist, grassy field or meadow. In British Columbia, however, it is a butterfly of the North, and the records of its occurrence trace almost perfectly the Alaska Highway. Thus, it is not the super-familiar species here that it is elsewhere. The Red-disked Alpine is the first alpine of spring, and it shows the typical bobbing flight style of the genus, flying low to the ground, where its red "disks" are easy to see as it flits past. Knowing a butterfly in different parts of its geographic range can give such different impressions of its nature. Red-disked Alpines are also found in the high mountains of central Asia. Ian Sheldon tells me that he even spotted one on the Alberta prairies "...near Sandy Point, on a bitterly cold, windy, spring day in June... I saw one fidgeting around in the grasses and on closer inspection, realized that there was one solitary ant maneuvering the [dead] critter through the hideous tangle of grass blades. What a chore!"

Similar Species

Mt. McKinley Alpine

Magdalena Alpine

ID: a typical, mid-sized, brown alpine, but with no eyespots, and a reddish mid-wing patch on the upper forewing; the outer half of the hind wing underside is greyish; sexes are similar.

Similar Species: *Mt. McKinley Alpine* (p. 294): shows no grey mottling on the underside. *Magdalena Alpine* (p. 293): all black.

Caterpillar Food Plants: grasses (Poaceae, *Poa* spp.), possibly sedges (Cyperaceae: *Carex* spp.).

Habitat & Flight Season: open areas in forests, valley bottoms in mountains; flies early in the season, primarily in May and June.

Magdalena Alpine
Erebia magdalena

Wingspan: about 44–50 mm

I have already mentioned this species in the context of lepidopterist Robert M. Pyle's writings, but there is more to it than its legendary status. In British Columbia, these butterflies live in an isolated population, near McBride, high in the Rockies, where they sip nectar at phlox and forget-me-not blossoms. And although we think of them as black, they actually have a lovely green-purple iridescence when they first emerge from the pupa. The adults often bask on dark, lichen-covered rocks on ridges because these are the warmest microhabitats available in such cool, alpine areas. They lay eggs on these rocks, and the newly hatched caterpillars have to wander in search of grasses to eat. To most butterfly experts, the Magdalena Alpine is tightly associated with Robert M. Pyle's name, but another friend of mine, Gerry Hilchie, has done most of the important fieldwork on this butterfly. In a seminar he gave on the subject, some time in the 1970s, a young and somewhat nervous Gerry Hilchie showed us a slide of a Magdalena egg, and asked the unintentionally hilarious rhetorical question "what could possibly be the advantages of getting laid on a rock?" Everyone laughed, the tension eased and Gerry gave a good talk.

Similar Species

Mt. McKinley Alpine

Also Called: Rockslide Alpine.

ID: an all-black alpine, and in fact our only all-black butterfly in B.C.; sexes are similar.

Similar Species: *Mt. McKinley Alpine* (p. 294): is almost all black, but still shows faint reddish patches in the forewings on the topside.

Caterpillar Food Plants: grasses (Poaceae).

Habitat & Flight Season: high altitude alpine rockslides and boulder fields; flies in July.

Mt. McKinley Alpine
Erebia mackinleyensis
Wingspan: about 45–50 mm

♀ ♂

Similar Species

Magdalena Alpine

Red-disked Alpine

The Mt. McKinley Alpine has long been considered a subspecies of the Magdalena Alpine, but we now treat it as separate, based on lepidopterist Gerry Hilchie's research. Like the Magdalena, this species is only known at one locality in British Columbia, found at Summit Lake. Ross Layberry and the other authors of *The Butterflies of Canada* point out that butterflies of the high alpine zone in the Rockies are typically different in appearance in the north and the south, and intermediate in between, in a gradual "cline." Thus, when Gerry Hilchie showed that the break between the Magdalena and Mt. McKinley Alpines was "clean," with no tendency in either species to resemble the other where they meet, this was enough to consider them as two separate species. This makes more sense when you think of the Madgalena Alpine as a North American butterfly, and the Mt. McKinley Alpine as a Siberian species, which also spills over the Bering region into the northern Rockies.

ID: almost all black, with a slight reddish flush on the forewing on the topside; sexes are similar.

Similar Species: *Magdalena Alpine* (p. 293): all black. *Red-disked Alpine* (p. 292): hoary on the hind wing underside.

Caterpillar Food Plants: unknown, but probably grasses (Poaceae) and sedges (Cyperaceae: *Carex* spp.).

Habitat & Flight Season: like the Magdalena, found on boulders and rocky slopes in the high alpine; flies in June and July.

Mountain Alpine
Erebia pawlowskii
Wingspan: about 30–40 mm

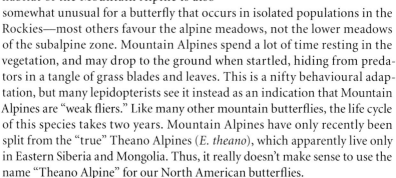

♂
♀

The Mountain Alpine is probably the most distinctive alpine of all. Unfortunately, most of us rarely if ever see it because the British Columbia range of this species consists of one locality alone—Stone Mountain Provincial Park (140 km west of Fort Nelson along the Alaska Highway). The habitat of the Mountain Alpine is also somewhat unusual for a butterfly that occurs in isolated populations in the Rockies—most others favour the alpine meadows, not the lower meadows of the subalpine zone. Mountain Alpines spend a lot of time resting in the vegetation, and may drop to the ground when startled, hiding from predators in a tangle of grass blades and leaves. This is a nifty behavioural adaptation, but many lepidopterists see it instead as an indication that Mountain Alpines are "weak fliers." Like many other mountain butterflies, the life cycle of this species takes two years. Mountain Alpines have only recently been split from the "true" Theano Alpines (*E. theano*), which apparently live only in Eastern Siberia and Mongolia. Thus, it really doesn't make sense to use the name "Theano Alpine" for our North American butterflies.

Also Called: Theano Alpine, Yellow-dotted Alpine or Banded Alpine; *E. theano*.

ID: the most distinctive of the alpines, with an upper mid-wing band of rectangular orange spots running through each wing; no eyespots within the orange; sexes are similar.

Similar Species: none, except at a distance, when it might be mistaken for another species of alpine.

Caterpillar Food Plants: unknown, but probably grasses (Poaceae) and sedges (Cyperaceae: *Carex* spp.).

Habitat & Flight Season: subalpine meadows and grassy bogs; flies mainly in July.

Arctics
(Genus *Oeneis*)

Macoun's Arctic on the floor of a pine and spruce forest

Arctics, like alpines, are a thrill to the deep butterfly devotee, and their mystique extends to (or more properly from) the Old World, where they are even more diverse than they are here. As with the alpines, a few species live in the sorts of places where people like to gather; others inhabit only the most inhospitable mountaintops and tundras. Some fly like grasshoppers, quickly flitting from one patch of ground to another, and others fly like nymphs or swallowtails, with powerful wingbeats interspersed with flat-winged glides. Arctics do not "bob" in flight. The colours of arctics are even more drab than those of alpines, but they are also more intricately patterned in earth tones and greys. With a good, close look, they are fairly easy to identify. The biggest obstacle to arctic appreciation is, however, the pronunciation of their genus name *Oeneis*. Let's face it, the English language has not prepared us for this word. Most lepidopterists I know say "EE-nee-ISS" but "EE-neez" is fine, too. Remember, there is no standard way to pronounce scientific names. The only real need is for you to feel confident in your conversations with others. Still, when combined with names such as *chryxus*, *polyxenes* and *taygete*, the whole thing becomes quite daunting, to the point where you might even wonder whether to pronounce a simple word like bore "bore" or "boh-ray." Do remember how to spell it, mind you, because spelling is indeed a standardized thing.

Great Arctic
Oeneis nevadensis
Wingspan: about 55–65 mm

The Great Arctic is exclusively a Pacific Northwest species, like the Vidler's Alpine. The subspecies *O. n. gigas*, the "Giant Arctic," is of "special concern" in British Columbia, and is the only subspecies known for certain in the province (although *O. n. nevadensis* is probably present in the southern interior near the Washington border). Thus, it is an uncommon butterfly, possibly because it prefers warmer habitats at lower elevations than do most of its close relatives. Butterfly enthusiasts count this as one of the most desirable of the satyrs to see or catch in the province. The males go to hilltops to watch for females, and they are quite aggressive in their pursuit of potential mates. Frequently, they land on tree trunks, fallen logs and on the bare ground of trails and dirt roads. Their two-year life cycle is true of all British Columbia arctics, as far as anyone knows.

Similar Species

Macoun's Arctic

Chryxus Arctic

Also Called: Nevada Arctic, Great Grayling, Felder's Arctic, Pacific Arctic or Giant Arctic.

ID: a large satyr, tan-coloured on the topside; scalloped wing margins; at least one eyespot on the underside of each wing; diffuse dark band through the hind wing underside; sexes are similar.

Similar Species: *Macoun's Arctic* (p. 298): female is less frosty near the front edge of the hind wing underside; has a more distinct band through the wing and darker wing bases on the topside. *Chryxus Arctic* (p. 299): smaller butterfly.

Caterpillar Food Plants: grasses (Poaceae).

Habitat & Flight Season: forest edges and clearings; flies in June and early July, but only in even-numbered years.

Macoun's Arctic
Oeneis macounii
Wingspan: about 55–65 mm

In many ways, this butterfly is much like its close relative, the Great Arctic. Male Macoun's Arctics, however, do not fly to hilltops as often to find females, and instead perch on logs and branches, and fly out from these lookout points near the ground. They often bask in the sun on paths and roads, leaning over to one side with both wings held above their back. Males lack the pheromone-dispersing scent scales, or "sex-scales" on the upper surface of the male forewing, but otherwise this could be a subspecies of the Great Arctic, at least in the opinion of many lepidopterists. The females, apart from very minor colour differences, are more or less identical between the two species. To the east of British Columbia, the range of the Macoun's Arctic follows the distribution of jack pine forests. In our province, it has broader habitat associations, to be sure. The Macoun's Arctic was named for John Macoun, a Canadian botanist.

Similar Species

Great Arctic

Chryxus Arctic

Also Called: Canada Arctic.

ID: another large arctic, tan-coloured on the topside, mottled on the underside; dark band through the underside of the hind wing; at least one eyespot on each wing; sexes are similar.

Similar Species: *Great Arctic* (p. 297): has darker wing bases on the topside, and a whiter leading edge to the hind wing underside. *Chryxus Arctic* (p. 299): smaller butterfly.

Caterpillar Food Plants: grasses (Poaceae) and sedges (Cyperaceae: *Carex* spp.).

Habitat & Flight Season: open, dry coniferous forests; flies in June, July and early August; appears in odd-numbered years in the Peace River region, but even-numbered years elsewhere.

Chryxus Arctic
Oeneis chryxus
Wingspan: about 40–55 mm

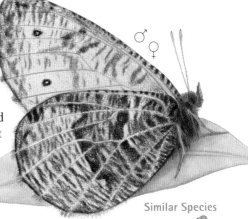

The Chryxus Arctic is the most widely distributed arctic in North America, occurring across the boreal forest and the Canadian Shield, and down the Rocky Mountains. It lives in a variety of habitats within this vast range, from grasslands to forest clearings to the treeless alpine zone. Wherever it is found, it is often seen sipping nectar from flowers, and several authors have suggested that yellow blossoms are especially to its liking. Mind you, some books say it is a slow flier, some say it is fast, and one even says that its flight is "non-directional." Frankly, if you can show me a butterfly with non-directional flight, I'll pay you back for the cost of this book. And what about the pronunciation of this horrible name? Chryxus is apparently a Greek word meaning "golden," and I have most often heard it pronounced "KRIKS-uss." Among other insect names, the Greek word for gold appears usually as "chrys" or "chryso," but at least the name Chryxus finds itself at home among the other consonant-rich arctic names (many of which were coined by Slavs, accustomed to such words).

Similar Species

Great Arctic

Macoun's Arctic

Also Called: Brown Arctic.

ID: another tan-coloured arctic on the topside, with one or two eyespots on each wing underside, or none; sexes are similar.

Similar Species: *Great Arctic* (p. 297) and *Macoun's Arctic* (p. 298): both are larger, but share the same general colour pattern; the *Great* is more like Chryxus dorsally, and *Macoun's* is similar ventrally.

Caterpillar Food Plants: grasses (Poaceae) and sedges (Cyperaceae: *Carex* spp.).

Habitat & Flight Season: meadows and open forests; flies in June, July and August, but earlier at lower elevations.

Uhler's Arctic
Oeneis uhleri
Wingspan: about 35–45 mm

In British Columbia, the Uhler's Arctic is a grassland species found only in the Peace River district, where its range is disjunct from other populations of the species on the open grasslands of the southern prairie provinces. These Peace River Uhler's Arctics are a bit larger and darker in colour than their southern cousins, and may well be named as a separate subspecies before long. Compared to the Chryxus Arctic, the Uhler's is generally slower in flight, and in many ways it is a lot like a grasshopper, flushing from cover, flying a short distance, and then dropping back into the grass stems. This is a behaviour pattern that meshes well with their mate-finding strategy. Uhler's Arctics prefer bunchgrass to other sorts of grasses, and the males wait patiently at the bases of bunchgrass clumps, then fly out to chase females. They are rarely found at flowers, perhaps because flowers are sometimes hard to come by in their preferred habitat during their flight season, and instead they often sip at mud puddles, or the dung of mammals, unbutterfly-like as that might seem. The original "Uhler," by the way, was a librarian at the Museum of Comparative Zoology at Harvard University in Massachussets. (We know he was male because "*uhleri*" ends in -i, not –ae.)

Also Called: Rocky Mountain Arctic.

ID: a relatively small, pale arctic; each wing usually has one or more eyespots; wiggly line through the forewing underside is relatively faint (or absent); sexes are similar.

Similar Species: *Alberta Arctic* (p. 301): the other small, pale grassland arctic has smaller and fewer eyespots, and a line through the forewing underside.

Caterpillar Food Plants: grasses (Poaceae).

Habitat & Flight Season: open grassy areas; flies in June and July.

Alberta Arctic
Oeneis alberta
Wingspan: about 35–45 mm

The Alberta Arctic is our smallest arctic, and the second species of small, grassland arctic restricted to the Peace River district within British Columbia. Like the Uhler's, it is also found in the grasslands of the Great Plains, and like Uhler's, it also flies like a grasshopper. The two species are a bit challenging to distinguish at first, but it helps that the Alberta is on the wing ealier in the season, so the first arctics you see are bound to belong to this species. Otherwise, the two species have parallel stories, even to the point that the Peace River populations of both may be new subspecies, larger and more contrastively marked than the same species on the prairies. As one travels farther and farther south over the range of the Alberta Arctic, it becomes a mountain species, restricted to the cool grassy tops of desert peaks in places like southern Arizona. How interesting that it is found in the warmest of places in Alberta, its namesake province, and the coolest of places near the Mexican border. One other note deserves mention here—it has been said that mating pairs of Alberta Arctics refuse to fly, and can be picked up by hand. I haven't noticed this myself, but now that I've heard the rumour, I'll be on the lookout.

Similar Species

Uhler's Arctic

Also Called: Prairie Arctic.

ID: a small arctic, with one or more small eyespots on the underside of each wing; distinct wiggly line through the forewing underside; sexes are similar.

Similar Species: *Uhler's Arctic* (p. 300): a bit larger, in general, and has more and larger eyespots and a less distinct line through the forewing underside.

Caterpillar Food Plants: grasses (Poaceae).

Habitat & Flight Season: grasslands; flies in May and early June.

Jutta Arctic
Oeneis jutta
Wingspan: about 35–55 mm

Jutta Arctics are forest butterflies, and the truth is that there are few species to which this distinction can be applied. Their flap-and-glide flight style hints at their membership in the brushfoot family of butterflies. Males, especially, are well known for returning to their territories after they have been disturbed, much the way a Red Admirable might. As well, they often rest on tree trunks, preferring upright, living trunks to fallen logs, although they will land on those too. They perch in wait for females, but they are also apparently able to detect newly emerged females by smell. In some places, their two-year life cycle makes them more common in odd-numbered years, while in others they are common in even-numbered years, or every year for that matter. This is another holarctic species, found also in the Old World, where it got its name. Jutland was an old Danish state, which is now part of Sweden. For that reason, and because I am part Swedish myself, I like to pronounce the name in an exaggerated Swedish accent, as the "YUUH-tah HARK-tik."

Also Called: Forest Arctic or Baltic Grayling.

ID: similar to a very pale alpine on the topside, with eyespots set in pale areas; sexes are similar.

Similar Species: relatively easy to recognize among the other arctics.

Caterpillar Food Plants: sedges and jointed rush (Cyperaceae: *Carex* spp. and *Juncus articulatus*), possibly also grasses (Poaceae).

Habitat & Flight Season: moist peatland forests and open pine forests; flies in June and July.

White-veined Arctic
Oeneis bore

Wingspan: about 40–50 mm

For those who abhor the ubiquitous taxonomic confusion that goes along with butterfly study, the name *bore* must seem especially apt. In truth, it means "north" in Latin, but I have to agree, it's a good, if unintentional, pun. *Oeneis bore* is apparently a species that occurs both in the Old World and the New, and there is no real consensus on whether the name *O. taygete* should be used for the North American members of this taxonomic entity. As with other holarctic butterflies, those who prefer to use one species name for the whole thing see the two isolated halves of the species as connected by history, not by some sort of shared essence. Those who prefer separate names for geographically separate butterflies look for what they consider significant differences of one sort or another and, in my opinion, sometimes think in essentialist ways. White-veined Arctics have a habit of flying uphill when startled, making them tough to catch at high altitude. As a corollary of this behaviour pattern, these butterflies prefer meadows just above treeline rather than the actual summits, above which there is no place to flee.

Similar Species

Polyxenes Arctic

Melissa Arctic

Philip's Arctic

Also Called: Arctic Grayling; *Oeneis taygete*.

ID: a large arctic with no eyespots; somewhat translucent wings; an obvious dark band through the underside of the hind wing; and the hind wing underside veins are lined with white; sexes are similar.

Similar Species: *Polyxenes Arctic* (p. 305) and *Philip's Arctic* (p. 306): lack the white veins. *Melissa Arctic* (p. 304): lacks the distinct dark band.

Caterpillar Food Plants: sedges (Cyperaceae: *Carex* spp.) and grasses (Poaceae).

Habitat & Flight Season: alpine tundra and forest meadows; flies in June and July.

Melissa Arctic
Oeneis melissa
Wingspan: about 35–50 mm

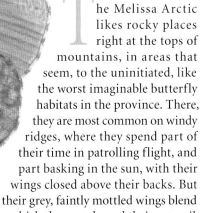

♂
♀

Similar Species

Polyxenes Arctic

The Melissa Arctic likes rocky places right at the tops of mountains, in areas that seem, to the uninitiated, like the worst imaginable butterfly habitats in the province. There, they are most common on windy ridges, where they spend part of their time in patrolling flight, and part basking in the sun, with their wings closed above their backs. But their grey, faintly mottled wings blend well with the rocks on which they perch, and their caterpillars find enough to eat to make life in the alpine zone worthwhile. Don Lafontaine, one of the authors of *The Butterflies of Canada*, has shown that, like underwing moths in the genus *Catocala*, Melissa Arctics prefer to land on surfaces that match their own colour. This species shows the opposite tendency to that of White-veined Arctics in that they fly downhill, not up, when they are spooked. This makes them dangerous to chase, at least for the butterfly-netting crowd, because Melissa Arctics often plummet over sheer cliffs as a means of escape. Perhaps the way to catch them is to send one person up top, and leave one down below, so they can scare the two sorts of arctics to each other (and yes, I'm at least partly joking here). Of the three Arctic arctics (Melissa, White-veined and Polyxenes), this is the most widespread in southern B.C.

Also Called: Mottled Arctic.

ID: a large, dark, somewhat translucent butterfly; no eyespots; no white wing veins; dark band on the underside of the hind wing is poorly defined. In other words, this is an extremely plain-looking bug; sexes are similar.

Similar Species: *Polyxenes Arctic* (p. 305): can be quite similar; occurs in the same areas; generally has a more distinct band through the hind underwing.

Caterpillar Food Plants: sedges (Cyperaceae: *Carex* spp.), possibly also grasses (Poaceae).

Habitat & Flight Season: alpine tundra, high in the mountains; flies in late June through early August.

Polyxenes Arctic

Oeneis polyxenes

Wingspan: about 35–50 mm

The Polyxenes Arctic, in western Canada at least, is the third of the high elevation trio of Polyxenes, Melissa and White-veined Arctics. All three have "thinly scaled" and "translucent" wings, and all three are about as subtly marked as a butterfly can be. Because they live in windy places, the males are more prone to perching and darting out at potential mates rather than patrolling and running the risk of being blown off the mountaintop. It's tough to get much of a feel for a butterfly's behaviour in these habitats, which probably accounts for the fact that some books refer to the Polyxenes Arctic as a fast flier while others describe it as slow or weak. Most of the time, it flies at a speed somewhere between slow and the speed of the wind on any given day. Having said that, I should add that those who know the species well say it is less prone to inhabit scree slopes and boulders, the preferred habitat of the Melissa Arctic. For lepidopterists from eastern North America, all of this may sound a bit odd because in the east, this species is most famous for its isolated population on the top of Mt. Katahdin, Maine. From a Rocky Mountain perspective, Maine is hardly the "rugged" place we picture when we think of Polyxenes Arctics. Because we may always find a butterfly, or any other organism for that matter, in one habitat, it doesn't mean that the species is unable to make a go of it elsewhere, if the chance arises.

Similar Species

Philip's Arctic

Also Called: Banded Arctic.

ID: this arctic has grey, faintly mottled wings; dark band through the underside of the hind wing is obvious; veins are brown, not white; sexes are similar.

Similar Species: *Melissa Arctic* (p. 304): is more grey, with a less defined hind wing band on the underside. *Philip's Arctic* (p. 306): more grey, less orange, but otherwise similar.

Caterpillar Food Plants: sedges (Cyperaceae: *Carex* spp.) and grasses (Poaceae).

Habitat & Flight Season: alpine zone; flies in July and August.

Philip's Arctic

Oeneis philipi

Wingspan: about 45–55 mm

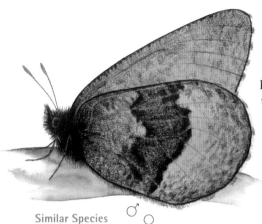

Similar Species

♂ ♀

Polyxenes Arctic

The Philip's Arctic is so far known only to inhabit Stone Mountain Provincial Park (140 km west of Fort Nelson along the Alaska Highway) in British Columbia, where it flies in peatland forests. In this regard, its behaviour is a lot like the Jutta Arctic—dodging between trees and landing on trunks and fallen logs. It is named for Kenelm Philip, the most prominent lepidopterist in Alaska. In this sense, it is unusual because almost all other patronymic butterflies are named for historic figures. I would hate to deny Kenelm his well-deserved namesake, and for that reason I find the name "Early Arctic" unappealing. The Europeans use the name "Rosov's Arctic" on their side of the pond (honouring the Russian, Vladimir Rosov). You may be wondering, however, how anyone can tell that this and the Polyxenes Arctics are separate species, given the fact that the only apparent difference between them is their habitats. Why not assume that the same species looks a bit different when it develops in different places? Well, if this proves eventually to be the case, it wouldn't surprise me a bit. But in the meantime, it is best to trust the opinions of those with the "big picture," familiar with both the species in the Canadian North as well as the one on the other side of the Bering Strait.

Also Called: Early Arctic or Rosov's Arctic; *O. rosovi*.

ID: dark grey-brown on the topside; and more or less identical to the Polyxenes Arctic; sexes are similar.

Similar Species: *Polyxenes Arctic* (p. 305): slightly smaller and less grey; never appears in black spruce peatlands.

Caterpillar Food Plants: grasses (Poaceae, notably *Eriophorum* spp.).

Habitat & Flight Season: spruce bogs and peatland forests; flies in late June and early July.

Milkweed Butterflies
(Subfamily Danainae)

A Monarch on a milkweed, the quintessential butterfly scene

The milkweed butterflies are a distinctly tropical group, but not simply because they are more diverse in the tropics, which is true of almost all of our butterfly families and subfamilies. That we have milkweed butterflies here at all is the result of the migratory habits of our single species, the Monarch. Not all milkweed butterflies are migratory, and for that matter, not all Monarchs are migratory. Characteristically, though, members of this group feed on milkweed plants and, as a consequence, they are able to sequester the cardiac glycoside toxins that the plants produce. Monarchs and their caterpillars are therefore poisonous to predators, and advertise this with their orange and black warning colours. The very similar Viceroy (which is part of the distantly related admiral subfamily) is a mimic of the Monarch, but it turns out that, in both species, some individuals are quite poisonous while others are palatable. Thus, they really mimic each other, mutually, as part of what we call a mimicry "ring," involving other species as one moves south toward the tropics.

Monarch

Danaus plexippus

Wingspan: about 90–100 mm

To the average person, almost any medium-to-large-sized butterfly is a "Monarch," but I suspect that you know better than that. The Monarch is our most famous butterfly, but only because it is famous in the United States. Here, the Monarch is a relatively uncommon denizen of the southern Interior that migrates north each year from the United States. New migrants arrive in June, and the British Columbia generation develops in mid-summer, as black-, yellow-and-white-banded caterpillars, on milkweed plants. The adults that result then migrate south (our only species known for sure to do so) to overwintering groves in California, and possibly farther south to Mexico. Unfortunately, the overwintering sites in California have seen fewer and fewer Monarchs as the

Similar Species

Viceroy

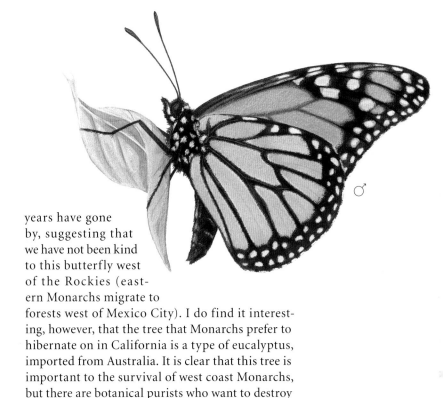

years have gone by, suggesting that we have not been kind to this butterfly west of the Rockies (eastern Monarchs migrate to forests west of Mexico City). I do find it interesting, however, that the tree that Monarchs prefer to hibernate on in California is a type of eucalyptus, imported from Australia. It is clear that this tree is important to the survival of west coast Monarchs, but there are botanical purists who want to destroy all of the eucalyptus groves in California because the tree is not native. Hopefully, they will not get their wish.

ID: large, orange butterfly with black wing veins, and a black forewing tip spotted with white. *Male:* has a dark swelling along the inner vein of the hind wing. *Female:* vein is smooth.

Similar Species: *Viceroy:* apparently extirpated in B.C., is smaller and has a dark line through the hind wing.

Caterpillar Food Plants: milkweeds (Asclepiadaceae: *Asclepias* spp.).

Habitat & Flight Season: found in fields and meadows in southern B.C.; flies from June through August.

Skippers
(Family Hesperiidae)

European Skipper (above); *Hobomok Skipper* (left); a species not found in British Columbia.

Skippers were so-named because of their skipping style of flight. But, you'll find that many of them do not fly in this fashion at all. Instead of watching for a particular flight style, you can recognize skippers by their relatively wide, somewhat rectangular heads. As well, most have very stout bodies. However, once you have the general look of a spread-winged skipper and a grass skipper in your head, none of this will be necessary. For a long time, books would refer to "butterflies *and* skippers," and some butterfly books included no mention of skippers at all. We now consider skippers to be the next closest relatives to the butterflies proper, and although the butterflies proper have been called "scudders" by some authors, this name has not caught hold. If you believe that skippers are more primitively mothlike than butterflies, then it also makes sense to put them before butterflies in the general order of things. If, on the other hand, you see butterflies and skippers as representing two different modifications of the mothlike form of their ancestors, then it is OK to place them at the end of the book, which is what I am doing here. Skipper caterpillars are also easy to recognize, and they typically have a narrow "neck" between the head and the thorax.

Plains Skipper

Spread-Winged Skippers
(Subfamily Pyrginae)

Grizzled Skipper

Spread-winged skippers, true to their name, usually perch with their wings spread out to the side. This makes them both easy to recognize, and easy to examine, at least on the dorsal surface. Having said that, I should also admit to you that some of the members of this group are our most difficult butterflies to identify. In fact, some are impossible without dissecting them. But try not to let this diminish your appreciation of spread-winged skippers. Focus instead on their lovely, hooked antennal tips, or the fact that they all feed on dicotyledonous plants. Look for them in other books, and admire these tropical butterflies for their incredibly colourful, iridescent and flashy looks. Then turn to the first member of our fauna, the Silver-spotted Skipper, and take some contentment from the fact that we have here an echo of the neotropics, and a living invitation to pursue these creatures far and wide. Notice as well that the spread-winged skippers generally have far fewer alternative names than do butterflies proper—one good thing about being a bit less conspicuous and less in the public eye.

Gilbert's Flasher, a Mexican spread-winged skipper, showing how colourful members of this group can be in the tropics

Silver-spotted Skipper
Epargyreus clarus
Wingspan: about 45–50 mm

This is one of the most widely distributed skippers in North America. Most of the time, you see these skippers in the morning, when males perch on shrubs and trees (almost never on the ground) to await passing females. With their wings open to the sun, they bask to warm up and fly out after any passing butterfly or bird, testing to see if it is a female of the species. During the heat of the day, they sometimes hang from the undersides of leaves. Their size and their speedy flight make them recognizable even when all you see is a brief blur across your path. Personally, I really like Silver-spotted Skippers, and I like the fact that their flight season is so wonderfully long. They spend the winter as a pupa, but apparently emerge over the course of the entire first half of summer. I also enjoy looking for the larvae on wild licorice. My son, Jesse, was the one who showed me how to find them when he was six years old, and a lot closer to the food plants than I am. Small larvae cut leaves and fold them over with silk to form a day-shelter, while big larvae spin many leaves together. They come out to feed at night.

ID: our largest skipper; long forewings; gold bar on the forewings; silver blotch on the underside of the hind wings.

Similar Species: unique among our skippers.

Caterpillar Food Plants: legumes, including wild licorice (Fabaceae: *Glycyrrhiza lepidota*).

Habitat & Flight Season: along forest edges, in some gardens and along streams; flies from mid-May to early August; one long flight season during which emergence is continual.

Northern Cloudywing

Thorybes pylades

Wingspan: about 35–45 mm

♂ ♀

The Northern Cloudywing's range extends all the way down to Mexico, so it is not really "northern" in the usual sense of the word. Where it occurs in the United States, however, it is part of a confusing group of very similar species. I find it interesting how this affects the psychology of butterfly watchers. We look at the Northern Cloudywing in Canada, and we say it is distinctive. Then we look at the duskywings and say they are confusing. But neither group of butterflies is more, or less, intrinsically easy to recognize than the other—it all depends on context. Should we be glad that there is only one cloudywing here, and that it is therefore easy to identify? Or should we be envious of our neighbors to the south, with greater cloudywing biodiversity? Northern Cloudywings generally perch on the ground, or on very low plants. They seem to like muddy ground as well, although they also sip at flowers. Usually, when you encounter one, you encounter only one—and they are never common in any single location.

ID: a relatively large, dark skipper, with just a few small white dots on its forewings.

Similar Species: a bit like the duskywings, but with a simpler pattern and more pointed wings; sexes are similar.

Caterpillar Food Plants: various herbaceous legumes (Fabaceae).

Habitat & Flight Season: meadows, clearings and along streams; flies from mid-April through July.

Dreamy Duskywing

Erynnis icelus

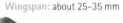

There are 17 species of duskywings in North America, and more in the Old World, so we are fortunate that this species is relatively easy to identify, albeit only because of the absence of a row of tiny spots that leave just a trace of themselves in the scale pattern of this species. The Dreamy Duskywing is a good introduction to the group, at least in the sense that it typically perches on the ground, or low to the ground. Its caterpillars are also typical of the group, in that they construct leaf nests much like those of the Silver-spotted Skipper. Unfortunately, it is not our most common species although it does range across Canada and southwest through the mountains. The name "dreamy" is somewhat poetic, but it has little to do with the butterfly itself. Instead, it is part of the long tradition of naming members of this genus after figures in Greek mythology. Icelus was the son of the god of sleep (and yes, there is also a Sleepy Duskywing farther south and east, *E. brizo*, with a similar story).

Also Called: Aspen Duskywing.

ID: a medium-sized spread-winged skipper; subtle blotches of greyish, purplish and brownish patches, with no translucent spots about three-quarters of the way out along the leading edge of the forewing. Male has more hoary white than the female.

Similar Species: the other duskywings are very similar, but all of our other species have the translucent spots.

Caterpillar Food Plants: willows and poplars (Salicaceae: *Salix* spp., *Populus* spp.).

Habitat & Flight Season: clearings and willowy areas; flies from late April through July.

Propertius Duskywing
Erynnis propertius
Wingspan: about 35–45 mm

♀

The Propertius Duskywing has been called the most common duskywing in its range, and at the very least this tells us that it is common where it occurs, in Garry oak woods (and among other oaks farther south). Personally, I think the name "Western Oak Duskywing" is far better than "Propertius"; the word Propertius, which makes reference to a Roman poet, is too easy to confuse with Persius and Pacuvius. Not that I have anything against Roman poets—I simply feel that these names (along with the names of the Frigga and Freija Fritillaries) constitute a mnemonic challenge we don't need. Males of this skipper species typically perch on or near hilltops, and also visit mud and flowers. As duskywings go, this species is probably the second easiest to identify in the province, next to the Dreamy Duskywing, but it has a certain look to it that makes it somewhat recognizable even when you don't see the light spots on the underside of the hind wing.

Similar Species

Dreamy Duskywing

Also Called: Western Oak Duskywing.

ID: a typical duskywing, with a row of translucent spots near the leading edge of the forewing (three quarters of the way toward the tip); two light spots near the leading edge of the hind wing underside; male has more tiny white hairs on the wings and looks more frosty than the female.

Similar Species: *Dreamy Duskywing* (p. 315): lack the translucent spots. The other three lack the pale spots on the hind wing.

Caterpillar Food Plants: Garry oak (Fagaceae: *Quercus garryana*).

Habitat & Flight Season: always in Garry oak woodlands; flies from late April, trailing into July.

Pacuvius Duskywing
Erynnis pacuvius
Wingspan: about 35–40 mm

Various books will tell you that this or that feature will allow you to recognize the Pacuvius Duskywing, but in practice, I honestly don't think I can distinguish it. You can, however, rule out the Afranius Duskywing, but not the Persius, because the Afranius' range does not overlap with the range of the Pacuvius (and north of the Pacuvius' range, only the Persius is likely to be encountered). Specialists suggest that dissection of the male genitalia is the way to tell these species apart, with the females identified by association with the males. You might, out of perverse curiosity, wonder what is involved in dissecting the genitalia. It requires a good dissecting microscope, some potassium hydroxide, fine forceps and other micro-dissection tools, glycerine, patience and good illustrations of what to look for. The book by Ehrlich and Ehrlich, *How to Know the Butterflies*, has instructions, and Guppy and Shepard's book, *Butterflies of British Columbia*, shows you what to look for. Try it if you want—after all, what's the worst thing that can happen?

Similar Species

Persius Duskywing

Afranius Duskywing

Also Called: Dyar's Duskywing or Buckthorn Dusky Wing.

ID: one of three species of duskywings in the province; has a row of translucent spots, perpendicular to the leading edge of the forewing, out near the tip; light spots near the base of the hindwing underside; sexes are similar.

Similar Species: *Persius Duskywing* (p. 319): and *Afranius Duskywing* (p. 318): are both almost impossible to distinguish from this species.

Caterpillar Food Plants: ceanothus (Rhamnaceae: *Ceanothus* spp).

Habitat & Flight Season: clearings and open woodlands; flies in May and June.

Afranius Duskywing
Erynnis afranius
Wingspan: about 30–35 mm

♂ ♀

About the only thing that lepidopterists have identified to help separate this species from the Persius Duskywing (apart from the shape of its male genitalia) is the fact that Afranius Duskywings do not hilltop, and Persius males do. Wouldn't it be simpler if Afranius were still considered a subspecies of Persius? Of course, I value truth above convenience, but recent speculation suggests that the Afranius might, in fact, be a subspecies of the Columbine Duskywing (*E. lucilius*), which in turn is almost indistinguishable from the Wild Indigo Duskywing (*E. baptisiae*), the species that surely possesses the most exciting-sounding English name in the group. It has been suggested that our one and only known population was accidentally introduced because this butterfly is found primarily east of the Rockies, and then as a disjunct anywhere near New Aiyansh, as well as in a few isolated populations in the Yukon.

Similar Species

Pacuvius Duskywing

Persius Duskywing

Also Called: Bald Duskywing.

ID: the second of three virtually identical duskywings; sexes are similar.

Similar Species: *Pacuvius Duskywing* (p. 317) and *Persius Duskywing* (p. 319).

Caterpillar Food Plants: legumes (Fabaceae).

Habitat & Flight Season: on the edge of a pasture at New Aiyansh, near the northern coast; flies from May through August.

Persius Duskywing
Erynnis persius
Wingspan: about 27–35 mm

When it comes to identification, at least this duskywing is common. If you see a duskywing that might be a Persius, and you call it a Persius, no one can fault you for it. And if you find one on a hilltop, you can be pretty sure it isn't an Afranius Duskywing. But in the southern part of the province, the Persius–Pacuvius problem reigns supreme, and is the cause of a certain amount of basic unhappiness. And if you find that this part of the book a bit discouraging, please understand me—that is not my intent. In fact, I may be preparing you for a much more confusing time to come. It is possible that in the near future more and more "cryptic species" will be identified by molecular techniques, as well as more "phylogenetic species" that may or may not correspond to what we now call subspecies. My point is that there is no reason to expect that nature will make all species easy to recognize, and no reason to expect that all biologists will even agree on what a species really is. Those who think otherwise, in the context of butterfly watching, are simply taking themselves too seriously.

Similar Species
Pacuvius Duskywing

Afranius Duskywing

Also Called: Hairy Duskywing.

ID: the third nearly identical duskywing in our fauna; sexes are similar.

Similar Species: *Afranius Duskywing* (p. 318) and *Pacuvius Duskywing* (p. 317).

Caterpillar Food Plants: lupines (Fabaceae: *Lupinus* spp.).

Habitat & Flight Season: meadows and stream edges; flies mostly in May, June and July.

Grizzled Skipper

Pyrgus centaureae

Wingspan: about 25–30 mm

This northern skipper just barely makes it into the United States. In British Columbia, this is the most familiar member of the genus *Pyrgus*, even though the Common Checkered Skipper holds this position almost everywhere else in Canada and the U.S. All members of the group are fond of perching on the ground, and the genus as a whole is mostly tropical. Lepidopterist Robert M. Pyle has made a coherent argument for changing the so-called official name to Alpine Skipper, even though these butterflies are not restricted to the alpine zone at the tops of high mountains. Britain's "Grizzled Skipper" is *P. malvae*, and both species overlap in areas of Eurasia because our Grizzled Skipper is holarctic. In North America, we have long ignored our naturalist counterparts in English-speaking Europe, but we really should try to standardize English names throughout the English-speaking world. Perhaps we should ask them to revert to the original name for their Grizzled Skipper, which was "The Gristle."

Similar Species

Common Checkered Skipper

Also Called: Alpine Checkered Skipper.

ID: a dark grey skipper with small white spots, a checkered wing fringe and a white hoariness overall; sexes are similar.

Similar Species: *Common Checkered Skipper* (p. 322): much more white. *Two-banded Checkered Skipper* (p. 321): much darker, and somewhat smaller.

Caterpillar Food Plants: possibly blackberries and their relatives, strawberries, or both (Rosaceae): *Rubus* spp.; *Fragaria* spp.).

Habitat & Flight Season: open areas; flies June through early August.

Two-banded Checkered Skipper

Pyrgus ruralis

Wingspan: about 23–28 mm

This butterfly is uncommon. It is also one of the first species on the wing in spring (at least of the butterflies that emerge from the pupa in spring). Thus, it may be more common than we think because most people do not enter their butterfly-chasing mindset until summer is well underway. Having said that, even those who venture out early in the spring are still likely to have a tough time finding the Two-banded Checkered Skipper, and it is likely that this will remain the most desirable of the three in British Columbia when it comes to butterfly watchers' wish lists. The two bands are subtle, wavy bands of white spots on the front wings of this butterfly. They are said to form a vague sort of "X" marking, but this is hard to see because there are also spots that are not part of the X. The X can be seen on our other two species of *Pyrgus* as well, with a little imagination. Study the pictures carefully, until you are clear on what constitutes a valid "X" and what does not.

Similar Species: Common Checkered Skipper

ID: a small black spreadwing skipper with white spots on the wings; sexes are similar.

Similar Species: smaller and darker than the *Grizzled Skipper* (p. 320), but with relatively larger white spots; darker and much smaller than the *Common Checkered Skipper* (p. 322).

Caterpillar Food Plants: Strawberries (Rosaceae: *Fragaria* spp.)

Habitat & Flight Season: fields, clearings and alpine meadows; flies from April through July.

Common Checkered Skipper

Pyrgus communis

Wingspan: about 25–32 mm

♂

The Common Checkered Skipper is a rare butterfly in British Columbia; in fact, conservationists consider it "of special concern." Elsewhere in North America (and south to Argentina!) it is often an extremely common and familiar butterfly. Lepidopterists have found, however, that the Common Checkered Skipper and

Similar Species

Grizzled Skipper

Two-banded Checkered Skipper

its closest relatives lack some of the features that characterize the northern members of the genus *Pyrgus*. They have suggested that this justifies proposing another genus for these species. If you look at butterfly classification, you will find that there are generally more genus names for skippers than there are for other butterfly groups. Skipper specialists seem to like many small genera, while those who study butterflies proper seem to prefer the opposite. I'm happy with the decision either way. I happen to be writing these words in south Texas, where the Common Checkered Skipper is indistinguishable from the White Checkered Skipper (*P. albescens*), and very similar to the Tropical Checkered Skipper (*P. oileus*). All I can say is that I look forward to getting back to Canada, where things are less confusing, even if the butterflies themselves are harder to find!

Also Called: Checkered Skipper.

ID: a grey and white skipper that is usually at least half white; female is grey and white or brown and white; male has light hairs on the wings that can make it appear blue when it is freshly emerged.

Similar Species: *Grizzled Skipper* (p. 320) and *Two-banded Checkered Skipper* (p. 321).

Caterpillar Food Plants: mallow family plants (Malvaceae).

Habitat & Flight Season: prefers dry, open places; flies in two broods, in May through July, and spottily in August and September.

Common Sootywing

Pholisora catullus

Wingspan: about 25–30 mm

The northern border of this species' range almost parallels the U.S.-Canada border. Thus, this is a skipper that is considered common and uninteresting by most American butterfly watchers, but distinctive and wonderful by those of us in the north. This is a truly "cute" butterfly, which almost always perches on or near the ground. I saw my first Common Sootywing near the Mexican border in Arizona, and the image of that particular butterfly pops back to me every time I see another. That is how distinctive it is. And, of course, the southern Interior of British Columbia reminds many people of the Sonoran Desert, so the presence of this butterfly serves as yet another reminder of our own "pocket desert," even though the Common Sootywing's range extends beyond the southern Okanagan by quite some distance.

Similar Species

Common Roadside Skipper

Also Called: Roadside Rambler.

ID: a small, spread-winged skipper that appears almost black, with a few tiny white dots on the forewing; some white on the head; sexes are similar.

Similar Species: *Common Roadside Skipper* (p. 343): somewhat similar, but with a very different overall shape.

Caterpillar Food Plants: lamb's quarters (Chenopodiaceae: *Chenopodium album*).

Habitat & Flight Season: open dry places; flies in two broods in late April through August.

Grass Skippers
(Subfamilies Hesperiinae and Heteropterinae)

Arctic Skipper (above); Woodland Skipper (right)

These skippers are generally smaller than the spread-winged skippers, and rarely if ever hold the wings out flat to the sides. Instead, they hold them closed over the back, out at about a 45-degree angle, or with the front wings at a 45-degree angle and the hind wings out flat. They are very easy to recognize as a group. One species in our fauna, the Arctic Skipper, is technically a member of the subfamily Heteropterinae. The members of this group have been called "intermediate skippers" in some books, and "skipperlings" in others. The features that distinguish the subfamily are difficult to see in the field, however, so I am including it here with the more diverse grass skippers proper, which belong to the subfamily Hesperiinae (some technical authors combine the two subfamilies into the Hesperiinae as well). The term skipperling does not imply that any such species is smaller than a typical skipper. I dislike the term skipperling, at least when referring to our fauna. In the southern United States, there are two truly tiny skippers (the Southern Skipperling and the Orange Skipperling, *Copaeodes minima* and *C. aurantiaca*), and in my opinion the name skipperling should be reserved for them.

Males in the grass skipper group usually have a different colour pattern than the females, and also show a black "brand" on the upper side of the forewing. For example, see the Draco Skipper on p. 337. This is a patch of scent-dispersing scales, also called "sex scales," "androconial scales" or the "stigma." The word "stigma" is the preferred term when discussing skippers. But keep in mind that when we speak of certain diseases carrying with them a "stigma," or "stigmatizing" their victims, it sounds as if the word stigma carries with it a negative connotation. This is certainly not the case when we are talking about skippers, and as in the original meaning of the word, it means simply "marked."

Arctic Skipper
Carterocephalus palaemon

Wingspan: about 20–30 mm

As mentioned above, this is the odd one out among the grass skippers, and is often placed in a separate subfamily. It is also found in Europe, where it is called the Chequered Skipper—an unfortunate overlap of names with our North American *Pyrgus* skippers. In England, the species was extirpated in 1976, but it is still found in Scotland. Is there a lesson here for us that we might someday lose this common butterfly as well? Perhaps, but it is more likely that the English situation is uniquely English. Here, the *mandan* race is considered of special concern, and is yet another example of a prairie province species that is also present in the Peace River district. This population, however, is not disjunct from other populations of Arctic Skippers in Alberta, and is therefore not one of the distinct subspecies found in the Peace. Lepidopterists John H. and Anna Comstock called this "a Canadian species," but almost every butterfly author has pointed out that it is not at all an arctic creature. Perhaps to our American colleagues, all of Canada is "Arctic," but to me it is sad that the name "Checkered Skipper" has been taken by another species on this side of the Atlantic. I personally really like Arctic Skippers, in part because they are recognizable at a glance, but also because they are handsome.

Also Called: Arctic Skipperling or Chequered Skipper.

ID: a small skipper; dark brown with orange angular spots on the topside; white spots on an orange background on the hind wing underside; sexes are similar.

Similar Species: all other small orange and brown skippers have a different pattern.

Caterpillar Food Plants: grasses (Poaceae).

Habitat & Flight Season: forest clearings, stream margins and grassy areas; flies primarily in May through July.

Garita Skipper

Oarisma garita

Wingspan: about 20–25 mm

This subtly gold-coloured skipper is another of my personal favourites. It is still an uncommon sight in British Columbia, but according to lepidopterist Robert M. Pyle, it has expanded its range in recent decades, and may become more abundant in the future. For now, however, it is a skipper of the Peace River district, disjunct from other Alberta populations, as well as flying in the southeast corner of British Columbia. Lepidopterists Crispin Guppy and Jon Shepard expressed some concern that the Garita may be out-competed by the introduced European Skipper, but personally I doubt it. In Alberta, in places where the European is extremely common, the Garita is still going strong. These are similar butterflies, to the point where they are often confused with each other.

Similar Species

European Skipper

Also Called: Garita Skipperling or Western Skipperling.

ID: a small, brownish skipper with a wash of orange colour over the upper surfaces of the wings. Hind wing underneath is orangy, with ever so slightly white-lined wing veins. Sexes are similar.

Similar Species: *European Skipper* (p. 328): more orange on the topside; has no white lining on the wing veins on the underside.

Caterpillar Food Plants: grasses (Poaceae).

Habitat & Flight Season: grassy areas; flies in June and July.

European Skipper
Thymelicus lineola

Wingspan: about 20–25 mm

Over the past ten years, I have come to realize that the European Skipper is a living wake-up call for those who disdain "invasive species" on so-called conservationist grounds. This butterfly was accidentally introduced into Ontario in 1910, and it is easily transported in hay because it is the only one of our skippers that overwinters as an egg. It has spread westward since, and in some places it has become the most abundant butterfly in the fauna. Some think that it is the lack of parasites that allows European Skippers to become so common, and at times, elsewhere, they can be pests in timothy hay. I think that ecological differences between European Skippers and native grass skippers are behind the success of the European Skipper here, and its ability to find room for itself among so many close relatives. If it is to be believed that an invasive species reduces biodiversity by outcompeting native species, the European Skipper does not provide any evidence of that. It is just as easy to argue that native skippers have been unaffected by the European, and that we now have more butterflies on the wing than we ever could have without this cute little addition to our fauna. My advice is to welcome the European Skipper, and avoid the weak logic of what some biologists now call "nativism."

Similar Species

Garita Skipper

Also Called: European Skipperling or Essex Skipper.

ID: a small, bright orange skipper with dark wing margins on the topside and plain orange wings on the underside; sexes are similar.

Similar Species: *Garita Skipper* (p. 327): darker, and has white on the hind wing veins on the underside.

Caterpillar Food Plants: grasses (Poaceae).

Habitat & Flight Season: grassy areas, both natural and manmade; flies from June through August.

Common Branded Skipper

Hesperia comma

Wingspan: about 25–30 mm

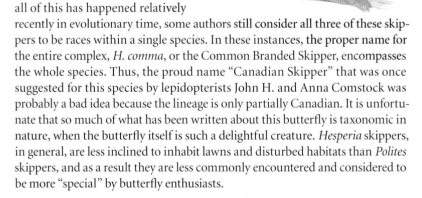

The Common Branded Skipper is the first of our *Hesperia* skippers, and the one you should learn best so that the others can be identified by comparison. It is also found in the Old World, and is yet another species that crossed the Bering region sometime long ago. It then gave rise to the Plains Skipper and the Western Branded Skipper, with the ancestral Common Branded lineage retaining the appearance of its Old World forebears. Because all of this has happened relatively recently in evolutionary time, some authors still consider all three of these skippers to be races within a single species. In these instances, the proper name for the entire complex, *H. comma*, or the Common Branded Skipper, encompasses the whole species. Thus, the proud name "Canadian Skipper" that was once suggested for this species by lepidopterists John H. and Anna Comstock was probably a bad idea because the lineage is only partially Canadian. It is unfortunate that so much of what has been written about this butterfly is taxonomic in nature, when the butterfly itself is such a delightful creature. *Hesperia* skippers, in general, are less inclined to inhabit lawns and disturbed habitats than *Polites* skippers, and as a result they are less commonly encountered and considered to be more "special" by butterfly enthusiasts.

Also Called: Holarctic Grass Skipper.

ID: a small orange and brown skipper with a more or less characteristic set of markings on the hind wing underside; male is more orange and has a dark stigma—the scent patch on the upper forewing.

Similar Species: The other *Hesperia* are all similar, especially the *Plains Skipper* (p. 330) and the *Western Branded Skipper* (p. 331).

Caterpillar Food Plants: grasses (Poaceae).

Habitat & Flight Season: grassy areas; flies primarily from mid-June through August.

Plains Skipper

Hesperia assiniboia

Wingspan: about 20–30 mm

This skipper is clearly related to the Common Branded, and is likely the prairie descendant of the European lineage of Common Branded Skippers that first made their way to North America across the Bering region. In British Columbia, Plains Skippers are found only in the Peace River district, in grassland areas. Like other Peace River disjuncts, they are under study to find out if they deserve the status of a distinct subspecies. Thus, the evolutionary history of this lineage may be complex indeed, and placing different names on lineages with separate histories is one way of highlighting their stories. Some authors have suggested that the Plains Skipper is more closely related to the Western Branded Skipper than to the Common Branded Skipper, but in truth they are all so close to one another that this may be nearly impossible to sort out. Of course, if all of this is too fiddly for you, you can always follow the current majority view and simply call them all Common Branded Skippers and be done with it.

Similar Species

Common Branded Skipper

Also Called: part of Common Branded Skipper; *H. comma assiniboia*.

ID: a typical *Hesperia* skipper; male is more orange and has a dark stigma.

Similar Species: the Plains Skipper looks like a washed out *Common Branded Skipper* (p. 329), and yes, older Common Branded Skippers do get "washed out" as well.

Caterpillar Food Plants: grasses (Poaceae).

Habitat & Flight Season: grassy areas of the Peace River district; flies mostly in August.

Western Branded Skipper

Hesperia colorado

Wingspan: about 25–30 mm

There is not much that can be said about this skipper than hasn't been covered under the preceding two close relatives. The story continues to build in complexity, however, because there are two distinct races of the Western Branded Skipper in British Columbia. In the southern Interior, one finds the *harpalus* race, and on southern Vancouver Island, the *oregonia* race. In other words, if this classification is followed, the pioneering lineage of Common Branded Skippers from Europe split into three species, after which the Western Branded Skipper split into component races itself. This is interesting. It tells us that evolution can happen quite quickly—at least fast enough to leave the "ancestral" type still in existence, and literally right next door to the descendant types. This is the sort of detective work that motivates butterfly study. In the midst of chasing and identifying butterflies, remember that these amazing creatures can tell us something about the history of life on earth.

Similar Species

Common Branded Skipper

Also Called: part of Common Branded Skipper, or Colorado Skipper; *H. comma colorado, H. c. harpalus* or *H. c. oregonia*.

ID: a typical *Hesperia* skipper; male is more orange and has a dark stigma.

Similar Species: *Common Branded Skipper* (p. 329): the most similar, but is a bit darker on the underside, and generally smaller.

Caterpillar Food Plants: grasses (Poaceae).

Habitat & Flight Season: dry grassy areas in the southern Interior, and on southern Vancouver Island; flies from July through mid-September.

Juba Skipper
Hesperia juba

Wingspan: about 30–35 mm

Finally, a *Hesperia* skipper that is relatively easy to recognize. If you are using other field guides, however, you might expect our Juba Skippers to be noticeably larger than their close relatives. My advice is to look for other characteristics, even though average Juba Skippers are larger than other *Hesperia*. The extra-large white spots on the hind underwing and the dark ground colour around the spots are the best things to watch for. You might also make note of the fact that Juba Skippers are not a hilltopping species, so in hilltop habitats, you are probably looking at other *Hesperia*. The name of the species comes from the Yuba River, in California. More interestingly, there is controversy over whether or not the adults of this species hibernate. Both sides of this mini-debate seem convinced, but the truth is that no one has yet found a hibernating adult, so no one knows for sure. If Juba Skippers do hibernate as adults, this would make them unique among the *Hesperia*.

Also Called: Yuba Skipper or Jagged-bordered Skipper.

ID: a crisply marked, orange and brown skipper on the topside, with especially large white patches on the greenish hind wing underside; male is more orange and has a dark stigma; the dark, deeply indented wing margin is clearly delineated from the orange ground colour, especially on the male.

Similar Species: generally a bit larger than its close relatives in the *Hesperia* genus.

Caterpillar Food Plants: grasses (Poaceae).

Habitat & Flight Season: grassy places among sage, or in open forests; flies from mid-May through September.

Nevada Skipper
Hesperia nevada

Wingspan: about 25–30 mm

The Nevada Skipper is a true hilltop butterfly, found almost always on dry hills and ridges. It looks one heck of a lot like the other *Hesperia*, but it is biologically quite distinct. The hind wing upperside is indeed different from that of other *Hesperia* skippers, but you really need to learn to judge the *degree* of crispness of the pale spots to use this feature effectively. Other people apparently "see" different things on this butterfly. For example, Guppy and Shepard say "the most reliable characteristic is the postmedian band in the ventral hind wing, which appears to be a connected series of irregular rectangles" (*Butterflies of British Columbia*, p. 107). I'm sure these men see what they claim they see, but when I look at the same markings I see crescents, triangles, swooshes and splotches, not rectangles, and it takes imagination to see what I am supposed to see. Thus, the job of learning to identify butterflies is, at least in part, a matter of learning to percieve things the way field guide authors do when they look at complex wing patterns. In truth, there is only so much that any field guide can communicate—the rest is up to you, and requires both patience and energy.

Also Called: Montane Skipper.

ID: much like the other *Hesperia* skippers, but the lowermost light spot on the hind wing underside is slightly offset toward the head; light spots on the hind wing underside are also clearly visible on the topside; male is more orange and has a dark stigma.

Similar Species: on other *Hesperia* species, the light spots on the hind wing underside cannot be seen as crisply on the topside; differences in habitat and flight season also help to distinguish the Nevada Skipper.

Caterpillar Food Plants: grasses (Poaceae).

Habitat & Flight Season: dry hilltops; flies in May and June.

Sachem Skipper

Atalopedes campestris

Wingspan: about 25–30 mm

The Sachem is a very familiar skipper throughout much of the United States, where it is common in gardens and disturbed habitats. Like many other butterflies, it often disperses northward in times of extreme abundance, and it is these "mass migrations" that have resulted in all of our British Columbia records for the species. It may be that Sachems fly south as well, but since their natural range extends well into South America, no one notices these movements, if indeed they really occur. In British Columbia, it appears that the species does not breed to form permanent populations, so all the Sachems we see here are one-way migrants that will die here without reproducing or returning to the south. The name of the Sachem means "chief," and refers to the large stigma of the male. I suppose that the male skipper with the biggest stigma probably smells more "mannish" than any other, and should therefore be granted chiefhood. How very macho and sexist of the person who named it! The genus *Atalopedes* is apparently very closely related to *Hesperia*.

Similar Species

Long Dash

Peck's Skipper

Also Called: Field Skipper.

ID: another small brown and orange skipper; male has a very large, oval stigma on the topside of the forewing, and a yellow patch on the underside of the hind wing; female has translucent spots on the topside, and an underwing pattern that is vaguely like that of the Long Dash.

Similar Species: *Long Dash* (p. 339): has similar pattern as that on the female's underwing. *Peck's Skipper* (p. 335): similar yellow patch on the underside of the hind wing. Both of these species should be easy to separate by referencing the illustrations.

Caterpillar Food Plants: grasses (Poaceae).

Habitat & Flight Season: grassy places; flies May through July.

Peck's Skipper

Polites peckius

Wingspan: about 20–30 mm

The Peck's Skipper is an easily recognized, familiar skipper over much of its range across central and eastern Canada, but in British Columbia, the Peck's Skipper is found only in the southeast corner and the Peace River district. As such, it doesn't have much chance to become a suburban species, the way it is elsewhere. The name of the species is, in my opinion, not as good as "Yellow-patch Skipper," which is descriptive of the primary field mark on this butterfly. To me, it looks ever so slightly like a miniature sketch of the map of North America. Professor Peck was a man who studied insects as agricultural pests in Cambridge, Massachusetts, back in the late 1700s.

Similar Species

Sachem Skipper

Also Called: Yellowpatch Skipper; *P. coras*.

ID: a small dark brown skipper; orange markings on the topside; a large yellow patch on the hind wing underside; male is more orange and has a dark stigma.

Similar Species: *Sachem Skipper* (p. 334): male is not as crisply marked on the underside, and has a distinctive large, rectangular stigma on the forewing topside.

Caterpillar Food Plants: grasses (Poaceae).

Habitat & Flight Season: grassy areas; flies in mid-June to mid-August.

Sandhill Skipper

Polites sabuleti

Wingspan: about 20-25 mm

The Sandhill Skipper is an interesting study in conservation thinking. In British Columbia this is a rare species of the southern Interior, but it is not strictly a butterfly of so-called natural habitats because it also frequents disturbed, early-successional habitats, and apparently benefits greatly from the nectar-bearing diffuse knapweed (*Centaurea diffusa*). Yet, despite its reliance on a "noxious" weed for nectar, it is considered a species of special concern in British Columbia, implying the need for some sort of conservationist protection. The Sandhill Skipper became established in British Columbia in the 1970s. It arrived on its own volition and is not an "invasive species." Examples such as the Sandhill Skipper help us to understand that biodiversity conservation is not necessarily a matter of fighting for "natural" habitats, or an unchanging, so-called balance of nature, and that human activities are always part of the overall picture, for better or for worse.

Also Called: Saltgrass Skipper.

ID: a variable, small skipper, but usually with the light areas of the hind wing underside "stretched out" along the wing veins; male is lighter on the topside and has a dark stigma.

Similar Species: the underwing pattern should distinguish this species from similar small skippers.

Caterpillar Food Plants: grasses (Poaceae).

Habitat & Flight Season: salt marshes, alkali flats and sagebrush flats; flies in two broods in mid-May to mid-June, and then late July through September.

Draco Skipper
Polites draco
Wingspan: about about 20–30 mm

On the east side of the Rocky Mountains, just a hop and a skip from British Columbia, the Draco Skipper is common and widespread. If the prevailing winds blew the opposite direction, this would surely be an equally common species in British Columbia. But it isn't, and instead it is found here only in the extreme northwest corner, near Atlin. There are, however, many populations in the Yukon Territory, and if you look at the overall range of the Draco Skipper, you will see that we just barely miss out on this species, thanks to the Rockies. It has been considered odd, however, that Draco Skippers do not occupy the Peace River grasslands, but these habitats are more like the Alberta parklands than the mountain meadows where the Draco usually lives, so I don't find the range of this species surprising at all. Because it occupies only a small corner of the province, it is of special concern in British Columbia. But this is clearly a byproduct of a biologically arbitrary political border and a biologically naïve system for ranking the conservation status of butterflies. And speaking of naïve, no one knows what entomologist W.H. Edwards was thinking when he named this butterfly after a dragon.

Also Called: Dragon Skipper or Rocky Mountain Skipper.

ID: a small brown skipper marked with orange, and with a mid-wing expansion of the outer light band on the underside of the very dark hind wing. Males are more orange and have a dark stigma.

Similar Species: similar to a *Hesperia* skipper, especially on the underside, but the details of its wing pattern are different.

Caterpillar Food Plants: grasses (Poaceae).

Habitat & Flight Season: found only in the Atlin area in the extreme northwest, despite large populations just east of the Rockies; flies in June and July.

Tawny-edged Skipper

Polites themistocles

Wingspan: about 20–30 mm

The Tawny-edged Skipper is more common in eastern Canada, but it is also reasonably common in British Columbia throughout its range in the southeast quarter of the province. At a distance, this butterfly looks quite dark, and on males, the orange patch may be visible even without binoculars. I am personally thankful for any skipper that is relatively easy to identify, but that is the not the only reason to appreciate the Tawny-edged Skipper. As a member of the genus *Polites*, the Tawny-edged does fairly well around people. It is a mid-summer species, and the scientific name refers to a Greek statesman, Themistocles. Thank goodness we don't have to call it the "Themistocles Skipper"; the pronunciation of Greek words is not a strong point for most North Americans. For example, in the southern U.S., there is a sulphur called the Lyside. I still remember sitting with lepidopterist Robert M. Pyle when he explained to a group of Texans that Lyside is a Greek word, pronounced "LISS-ih-DAY." The group looked at him, shrugged, and went on saying "LYE-side!"

ID: a small brown skipper with a distinct orange patch along the leading edge of the base of the forewing topside; a very plain dark hind wing underside; male shows more orange than the female.

Similar Species: among the members of a very uniform group, this one is more or less distinct.

Caterpillar Food Plants: grasses (Poaceae).

Habitat & Flight Season: moist, grassy areas; flies from late May through July.

Long Dash

Polites mystic
Wingspan: about 25-30 mm

The "long dash" on a Long Dash is made up of the male stigma and some dark markings that connect to it. This is a species of the Peace River grasslands and the southeast corner of British Columbia, but it is much more widespread and familiar to the east, across southern Canada and the northern United States. When you encounter this butterfly, be sure you are not looking at the more common Woodland Skipper. And don't feel bad about the trickiness of separating the two in the field—apparently, this pair of species shows parallel variation in wing pattern over their shared geographic range. In other words, the most Long Dash-like of the Woodland Skippers always occurs right alongside the most Woodland-like of the Long Dashes. Some greater fritillaries also show this pattern, and we interpret it as the result of similar evolutionary pressures on both species in the same environment.

♂

♀

Similar Species

Woodland Skipper

Also Called: Long Dash Skipper or Mystic Skipper.

ID: a small, brown and orange skipper with a yellow half circle on the hind wing underside, centered around a central yellow splotch; male is more orange, and has a dark stigma.

Similar Species: *Woodland Skipper* (p. 341): has a less distinct underwing pattern, and a more crisply indented, dark upperwing border.

Caterpillar Food Plants: grasses (Poaceae).

Habitat & Flight Season: wet meadows in the south, but drier places in the Peace; flies in June through August.

Sonoran Skipper

Polites sonora

Wingspan: about 25–30 mm

The Sonoran Skipper in British Columbia is a species of dry, grassy slopes in the Similkameen River basin, in the province's extreme south. This area may remind some people of the Sonoran region to the south, which includes more than just the Sonoran Desert, but is still based around dry grasslands and semi-deserts. But the Sonoran Skipper lives nowhere near anywhere that could reasonably be called "Sonoran," and was named in error. The Sonoran Skipper is "threatened" in British Columbia, but to me, the species is simply at the northern extent of its range in B.C., taking advantage of small patches of usable habitat. It is not at all under threat farther south. As climate changes, and habitats are altered, it is just as likely that this skipper will become more common as it is that the opposite will occur. Rarity is not necessarily equivalent to vulnerability; in fact, it may be a harbinger of potential future success.

Similar Species

Long Dash

Also Called: Western Long Dash.

ID: a small orange and brown skipper with a yellow half circle on the hind wing underside, centered around a central yellow splotch.

Similar Species: *Long Dash* (p. 339): has a similar underside pattern, but with thicker markings.

Caterpillar Food Plants: grasses (Poaceae).

Habitat & Flight Season: dry grassy slopes; flies in June and July.

Woodland Skipper

Ochlodes sylvanoides

Wingspan: about 25–30 mm

The Woodland Skipper is certainly one of our most abundant grass skippers, and it may well be the most common overall. Maybe this will always be the case, but maybe not—the European Skipper is also abundant when it finds itself in good habitat. Don't expect to see the Woodland Skipper only in woodlands, however! It was named for its resemblance to the European species *O. sylvanus*. *Sylvanus* means "of the forest," and *–oides* means "looks like." But none of our skippers are really forest species, and all prefer the feeling of sun on their wings and the sight of swaying grasses all around them. In Alberta, the Woodland Skipper is often the most abundant skipper on the southern prairies, but it has been recognized as a prairie species only within the last ten years because of its close resemblance to the Long Dash. Perhaps more British Columbian butterfly people should cross the mountains and help out their prairie colleagues from time to time, and vice versa!

Similar Species

Long Dash

Also Called: Western Skipper.

ID: an orange and brown skipper with deeply indented, dark wing borders on the topside, and the faintest hint of a light crescent on the hind wing underside; males are more orange and have a dark stigma.

Similar Species: *Long Dash* (p. 339): has smoother wing borders on the topside, and a more clearly defined pattern on the hind wing underside.

Caterpillar Food Plants: grasses (Poaceae).

Habitat & Flight Season: open, grassy areas; flies from July through September.

Dun Skipper

Euphyes vestris

Wingspan: about 25–30 mm

In British Columbia, the uncommon Dun Skipper is a butterfly of the Fraser River valley, the Lower Mainland and southeastern Vancouver Island. In general, it is a butterfly of almost all of eastern Canada and the Pacific Northwest coast. Only a few have been found in Alberta. It remains to be seen whether the disjunct eastern and western portions of this species' range are eventually given different names. And speaking of names, it seems that the male Dun Skipper was described as *E. ruricola*, and the female was named *E. vestris*, by the same author, in the same paper. Because the name *ruricola* came first in the paper, it should have become the official name, but it didn't. Apparently, Jean de Boisduval, the man who wrote the paper, labeled his male type specimen "*rubicola*," not "*ruricola*," and thereby muddied the waters even more. Because the rules of taxonomy are rather strict (and the type specimen needs to have the same name that appears in the formal description), these mistakes could not be overlooked, and necessitated the change to *vestris*.

Similar Species

Common Roadside Skipper

Tawny-edged Skipper

Also Called: Sun Sedge Skipper or Sedge Witch; *E. ruricola*.

ID: a remarkably plain brown, small skipper; female has a few white spots on the topside.

Similar Species: only vaguely like the *Common Roadside Skipper* (p. 343): or the *Tawny-edged Skipper* (p. 338), our only other truly dark, grass skippers.

Caterpillar Food Plants: the only grass skipper in our fauna that feeds on sedges (Cyperaceae).

Habitat & Flight Season: open areas near sedges; flies from mid-May to mid-August.

Common Roadside Skipper

Amblyscirtes vialis
Wingspan: about 20–25 mm

The Common Roadside Skipper is typically found low to the ground, in clearings in aspen poplar forests. Once you learn to recognize it, it becomes a familiar and reassuring sight. Common Roadside Skippers are known for their belligerence to other butterflies and their willingness to attack even birds, people and deer. In Alberta, when we first encountered this species in good numbers in the mid-1970s, we assumed that these skippers were experiencing an unusual population spike. Thirty years later, it's pretty clear that we simply weren't very good at finding them before, and that populations are relatively stable, at least for a skipper. This butterfly is our only member of a very large genus that is especially diverse and confusing in the southwestern United States. Once you have spent time in Arizona, attempting to tell one from the other, you will probably see much more in your average Common Roadside Skipper than you ever did before.

Similar Species

Common Sootywing

Also Called: Roadside Skipper or Black Little Skipper.

ID: a small, dark brown skipper with a few small white spots near the leading edge of the forewing, a checkered wing fringe, and a purplish dusting on the outside edges of the wings underneath; sexes are similar.

Similar Species: *Common Sootywing:* (p. 324): vaguely similar, but is clearly not a member of the grass skipper group.

Caterpillar Food Plants: grasses (Poaceae).

Habitat & Flight Season: deciduous forest clearings and, appropriately, along roadsides, in May, June and July.

Glossary

admiral band: a light band of colour that runs through the middle portion of both the front and hind wings.

androconium (pl. androconia): a patch of scales on a male butterfly's wing that gives off a chemical pheromone that aids in courtship. Also known as sex scales, or the stigma.

anterior: toward the head, or the front of the head.

aurora: a row of red or orange spots on the hind underwing of a member of the blues (Lycaenidae: Polyommatinae).

base: where a structure attaches to the body.

biocontrol: the use of predators and parasites, rather than poisons, to control pests.

borrow pit: a pit dug during construction to provide sand or gravel to other parts of the construction site.

bottom surface: on a butterfly's wings, the surface on which the hind wings overlap the front wings, and which faces downward when the butterfly's wings are fully spread to the sides.

butterfly season: the period during which butterflies of any sort are active and on the wing (and which is much shorter at high elevations and in northern areas).

buttergack: general term for the annoying petty politics that butterfly enthusiasts sometimes exhibit.

cell: an area on a butterfly wing, generally near the middle, outlined in veins and usually larger and more oval than other such areas on the wing.

cell bar: a dark, bar-shaped mark at the outer end of the cell, usually of the front wing.

cell-end bar/spot: a dark bar of coloured scales at the apex of the cell.

chrysalis (pl. chrysalides): the pupa of a butterfly.

cline: a gradual gradient in one or more features of a butterfly, expressed over a large geographic area.

cocoon: the silk covering that some butterfly caterpillars spin around their pupa.

colony: an isolated local population of a species of butterfly.

complete metamorphosis: a life history that passes through egg, larva (caterpillar), pupa and adult stages.

convergence: evolution of dissimilar creatures to become more similar owing to similar effects of the environment.

cryptic species: outwardly identical species that nonetheless do not interbreed and differ in ways that are not easy for people to detect.

dicotyledonous plants: broad-leaved deciduous plants.

discocellular spot: a light or dark spot in the hind wing cell of a sulphur butterfly.

disjunct: geographically separate from other members of the same species.

doubled: a pair of adjacent, very similar markings.

early successional: an environment during the period soon after a disturbance such as fire, construction, flood or landslide.

ecology: the study of the relationships between living things and their surrounding environment, both living and non-living.

eversible: capable of being turned inside out, or simply extended outward.

extirpated: locally extinct.

family: a group of species, in one or more genera, with a name that ends in *–idea* for animals and *–aceae* in plants.

forewings: the anterior pair of wings on a butterfly. Same as front wings, mesothoracic wings and primaries.

form: one of many colour morphs or phases in a butterfly population.

front wings: the anterior pair of wings on a butterfly. Same as forewings, mesothoracic wings and primaries.

genus: a group of species, with a name that is always capitalized and italicized.

girdle: a band of silk that supports some sorts of butterfly pupae around the middle of the pupa's body when attached to twigs.

ground colour: the predominant background colour on a butterfly wing.

hilltopping: the tendency of some male butterflies to congregate on the tops of hills and await passing females.

hind edge: the posterior or trailing edge of the wing.

hind wings: the posterior pair of wings on a butterfly. Same as metathoracic wings and secondaries.

hind wing angle: the free corner of the hind wing closest to the body of the butterfly.

holarctic: occurring in northern North America, northern Europe and Asia.

honey glands: glands on the caterpillars of some gossamer-winged butterflies, which produce a sugar solution that is eaten by ants.

inner surface: the upper or dorsal surface of a butterfly's wings, when the wings are held together over the body. This is the surface in which the forewings overlap the hind wings.

instar: the period between the growth of new skin (or exoskeleton) in a caterpillar. Note that the new skin begins growing beneath the existing skin, and that the period between sheddings of the skin is called the stadium.

intergrades: population within a species that is intermediate between two subspecies.

lateral margin: the outside edge of the wing (lateral means to the sides).

leading edge: the anterior edge of the wing, toward the head end of the butterfly.

lumper: someone who prefers fewer but more inclusive categories in taxonomy, as a result of emphasizing similarities.

marginal crescents: crescent-shaped marks along the outer margin of the wing.

mesothorax: the middle segment of the thorax, which bears the second pair of legs and the forewings.

metathorax: the third, hindmost segment of the thorax, which bears the third pair of legs and the hind wings.

mimicry ring: a group of species all of which share a common colour pattern in order to warn predators that they are distasteful, or to give the impression that they are distasteful when they are merely mimicking distasteful members of the ring.

neotropics: the warm tropical areas of Mexico, Central America and South America.

osmeteria: the smelly anti-predator glands that are everted from the neck region on the caterpillars of Parnassians and swallowtails.

patronymic: named after a man.

phylogenetic species: one of many species concepts, wherein any evolutionary lineage in which there is a newly evolved feature is considered a species (many other species concepts would treat some phylogenetic species as subspecies).

Pollard walk: a method for surveying butterfly populations, typically involving a weekly walk along a predetermined route, counting all butterflies seen within a certain distance of the observer.

posterior: toward the tail or rear of the animal.

post-median: to the rear, or to the outside, of the middle of the wing.

primitive: possessing characteristics in common with a distant ancestor.

pupa (pl. pupae): the resting stage of a butterfly's life cycle, during which it transforms from a caterpillar to an adult butterfly.

satellite spot: a small spot next to the larger discocellular spot on a sulphur butterfly's wings.

sequester: to retain a chemical in the body, unchanged from its original form.

sex scales: a patch of scales on a male butterfly's wing that gives off a chemical pheromone that aids in courtship. Also known as androconia, or the stigma.

sibling species: species that are difficult or impossible to distinguish by looking at them, and which may or may not be each other's closest relatives.

sp.: abbreviation for a single species.

species: a group of living things that interbreed freely in nature and are more or less distinct from other such groups (this is the so-called biological species concept, and the one used in this book).

species group: a group of closely related species, smaller than a subgenus.

sphragus: a glue-like plug produced by male Parnassians to prevent mated females from engaging in further copulations.

splitter: someone who prefers more and less inclusive categories in taxonomy, as a result of emphasizing differences.

spp.: abbreviation for more than one species.

ssp.: abbreviation for subspecies.

stadium (pl. stadia): periods of the life cycle in between sheddings of the skin (the exoskeleton).

stigma: a patch of scales on a male grass skipper's wing that give off a chemical pheromone that aids in courtship. Known in other butterflies as androconia, or sex scales.

structural colour: colour produced by the microscopic structure of butterfly wing scales, by diffraction or refraction of light, and not by pigmented chemicals in the scales.

subfamily: a major taxonomic division of a family, with a name that ends in *–inae*.

subgenus: a taxonomic subdivision of a genus, expressed by placing the subgenus name in parentheses following the genus name.

submarginal spots: refers to spot-like butterfly wing markings that are just inward from the outward (lateral) margin of the wing.

superfamily: a taxonomic grouping of a number of families, with a name that ends in *–oidea*.

synonymize: to make one taxonomic name equivalent to another, and thereby render one name obsolete.

taxon (pl. taxa): a word for a taxonomic group of any rank.

thecla spot: a spot on the hind wing undersides of hairstreak butterflies, along the lateral margin. This spot usually has a tail associated with it, and is thought to serve as a false eye to deflect the attacks of bird predators.

top surface: the upper or dorsal surface of a butterfly's wings. This is the surface in which the forewings overlap the hind wings.

trailing edge: the posterior or rearmost edge of a butterfly wing.

tribe: a taxonomic subdivision or rank, below that of superfamily and above genus.

tubules: very small tubes.

type locality: the place where the type specimen of a species was captured.

type species: the species that is designated as the one and only species in a genus that will always belong in that genus, unless the genus is synonymized. Other species can also belong to the genus, but if the genus is subdivided, the genus name always goes with the group that includes the type species.

type specimen: the specimen (or specimens) that are designated as certainly belonging to a given species at the time that the species is formally described in the scientific literature.

ultraviolet reflectances: patterns visible in a combination of visible and ultraviolet light, on a butterfly's wings.

underwing: the lower or ventral surface of a butterfly's wings. This is the surface in which the hind wings overlap the forewings.

upper surface: the top or dorsal surface of a butterfly's wings. This is the surface in which the forewings overlap the hind wings.

upper wing: the top or dorsal surface of a butterfly's wings. This is the surface in which the forewings overlap the hind wings. Same as upper surface.

wing bases: the area where the wings attach to the body.

wing margin: the outer or lateral edge of the wing.

Resources

Acorn, John. 1993. *Butterflies of Alberta.* Lone Pine Publishing, Edmonton.

Bartlett Wright, Amy. 1993. *Peterson First Guide to Caterpillars of North America.* Houghton Mifflin Co., Boston and New York.

Brock, Jim P., and Kenn Kaufman. 2003. *Butterflies of North America.* Kaufman Focus Guides. Hillstar Editions, Houghton Mifflin Co, New York.

Comstock, John A. 1883. *Butterflies of California.* Reprinted in 1989 by Scientific Publishers, Gainesville, Florida.

Comstock, John H. and Anna B. Comstock. 1904. *How to Know the Butterflies: A Manual of the Butterflies of the Eastern United States.* D. Appleton and Co., New York.

Ehrlich, Paul R. and Anne H. Ehrlich. 1961. *How to Know the Butterflies.* Wm. C. Brown Co. Publishers, Dubuque.

Emmel, Thomas C., ed. 1998. *Systematics of Western North American Butterflies.* Mariposa Press, Gainesville, Florida.

Glassberg, Jeffrey. 2001. *Butterflies Through Binoculars: The West.* Oxford University Press.

Guppy, Crispin S. and Jon H. Shepard. 2001. *Butterflies of British Columbia.* UBC Press, Vancouver, Toronto.

Holland, W.J. 1931. *The Butterfly Book.* Revised Edition. Doubleday and Company, Garden City, New York.

Howe, William H., ed. 1975. *The Butterflies of North America.* Doubleday and Company, Garden City, New York.

Layberry, Ross, Peter W. Hall, and J. Donald Lafontaine. 1998. *The Butterflies of Canada.* University of Toronto Press, Toronto.

Opler, Paul A., and Amy Bartlett Wright. 1999. *A Field Guide to Western Butterflies.* Peterson Field Guide Series. Houghton Mifflin Co., New York.

Opler, Paul A., and Andrew D. Warren. 2002. *Butterflies of North America. 2. Scientific Names List for Butterfly Species of North America, north of Mexico.* Contributions of the C. P. Gillette Museum of Arthropod Diversity, Colorado State University.

Pyle, Robert Michael. 1981. *The Audubon Society Field Guide to North American Butterflies.* Alfred A. Knopf, New York.

Pyle, Robert Michael. 1992. *Handbook for Butterfly Watchers.* Houghton Mifflin Co., Boston and New York.

Pyle, Robert Michael. 2002. *The Butterflies of Cascadia.* Seattle Audubon Society, Seattle.

Tilden, J.W., and Arthur C. Smith. 1986. *A Field Guide to Western Butterflies.* Peterson Field Guide Series. Houghton Mifflin C., Boston.

Weed, Clarence M. 1917. *Butterflies Worth Knowing.* Doubleday, Page and Co., Garden City, New York.

Weed, Clarence M. 1923. *Canadian Butterflies Worth Knowing.* The Musson Book Co., Toronto.

Societies

Entomological Society of British Columbia
Dr. Robb Bennett, Secretary Treasurer, B.C. Ministry of Forests
7380 Puckle Road, Sannichton, B.C. V8M 1W4
website: http://www.harbour.com/commorgs/ESBC/about.html
e-mail: Robb.Bennett@gems6.gov.bc.ca

Entomological Society of Canada
1320 Carling Avenue, Ottawa, Ontario K1Z 7K9
website: http://www.biology.ualberta.ca/esc.hp/homepage.htm

North American Butterfly Association
4 Delaware Road, Morristown New Jersey 07960
website: http://www.naba.org

Lepidopterists Society
c/o Los Angeles County Museum
900 Exposition Boulevard
Los Angeles, California 90007-4057
website: http://www.furman.edu/~snyder/snyder/lep/

Entomological Supplies

Bio Quip Inc.
17803 LaSalle Avenue
Gardena, California 90248-3602
phone: (310) 324-0620, fax: (310) 324-7931
e-mail: bioquip@aol.com.

Atelier Jean Paquet
3, rue du Côteau
Case Postale 953
Pont-Rouge, Québec G3H 2E1
website: http://www.quebecinsectes.com/pages/pages_english/macrodontia_english.html

Checklist

The following checklist will help you keep track of your butterfly watching experiences in British Columbia. Following each butterfly's scientific name is the last name of the person who first described the species—the "author" of the species. If the species has been placed in a different genus since it was first described, the author's name is placed in parentheses.

SWALLOWTAILS AND PARNASSIANS (Family Papilionidae)
Parnassians (Subfamily Parnassiinae)
- ☐ Rocky Mountain Parnassian, *Parnassius smintheus* Doubleday
- ☐ Clodius Parnassian, *Parnassius clodius* Ménétriés
- ☐ Eversmann's Parnassian, *Parnassius eversmanni* Ménétriés

Swallowtails Proper (Subfamily Papilioninae)
Old World Swallowtail Group
- ☐ Old World Swallowtail, *Papilio machaon* Linnaeus
- ☐ Anise Swallowtail, *Papilio zelicaon* Lucas
- ☐ Indra Swallowtail, *Papilio indra* Reakirt

Tiger Swallowtail Group
- ☐ Western Tiger Swallowtail, *Papilio rutulus* Lucas
- ☐ Canadian Tiger Swallowtail, *Papilio canadensis* Rothschild & Jordan
- ☐ Pale Swallowtail, *Papilio eurymedon* Lucas
- ☐ Two-tailed Swallowtail, *Papilio multicaudata* W.F. Kirby

Western Tiger Swallowtail

WHITES AND SULPHURS (Family Pieridae)
Whites, Marbles and Orangetips (Subfamily Pierinae)
- ☐ Pine White, *Neophasia menapia* (C. & R. Felder)
- ☐ Western White, *Pontia occidentalis* (Reakirt)
- ☐ Checkered White, *Pontia protodice* (Boisduval & LeConte)
- ☐ Becker's White, *Pontia beckerii* (W.H. Edwards)
- ☐ Spring White, *Pontia sisymbrii* (Boisduval)
- ☐ Veined White, *Pieris oleracea* Harris
- ☐ Margined White, *Pieris marginalis* Scudder
- ☐ Arctic White, *Pieris angelika* Eitschberger
- ☐ Cabbage White, *Pieris rapae* Linnaeus
- ☐ Large Marble, *Euchloe ausonides* (Lucas)
- ☐ Northern Marble, *Euchloe creusa* (Doubleday)
- ☐ Green Marble, *Euchloe naina* Kozhanchikov
- ☐ Desert Marble, *Euchloe lotta* Beutenmüller
- ☐ Sara Orangetip, *Anthocharis sara* Lucas
- ☐ Stella Orangetip, *Anthocharis stella* W.H. Edwards

Sulphurs (Subfamily Coliadinae)
- ☐ Clouded Sulphur, *Colias philodice* Godart
- ☐ Orange Sulphur, *Colias eurytheme* Boisduval
- ☐ Giant Sulphur, *Colias gigantea* Strecker
- ☐ Pink-edged Sulphur, *Colias interior* Scudder
- ☐ Pelidne Sulphur, *Colias pelidne* Boisduval & LeConte

- [ ] Palaeno Sulphur, *Colias palaeno* (Linnaeus)
- [ ] Western Sulphur, *Colias occidentalis* Scudder
- [ ] Alexandra's Sulphur, *Colias alexandra* W.H. Edwards
- [ ] Christina's Sulphur, *Colias christina* W.H. Edwards
- [ ] Mead's Sulphur, *Colias meadii* W.H. Edwards
- [ ] Hecla Sulphur, *Colias hecla* Lefébvre
- [ ] Canada Sulphur, *Colias canadensis* Ferris
- [ ] Nastes Sulphur, *Colias nastes* Boisduval

Sara Orangetip

GOSSAMER-WINGED BUTTERFLIES
(Family Lycaenidae)

Coppers (Subfamily Lycaeninae)
- [ ] American Copper, *Lycaena phlaeas* (Linnaeus)
- [ ] Lustrous Copper, *Lycaena cupreus* (W.H. Edwards)
- [ ] Bronze Copper, *Lycaena hyllus* (Cramer)
- [ ] Gray Copper, *Lycaena dione* (Scudder)
- [ ] Blue Copper, *Lycaena heteronea* Boisduval
- [ ] Dorcas Copper, *Lycaena dorcas* W. Kirby
- [ ] Purplish Copper, *Lycaena helloides* (Boisduval)
- [ ] Lilac-bordered Copper, *Lycaena nivalis* (Boisduval)
- [ ] Mariposa Copper, *Lycaena mariposa* (Reakirt)

Hairstreaks and Elfins (Subfamily Theclinae)
- [ ] Coral Hairstreak, *Satyrium titus* (Fabricius)
- [ ] Behr's Hairstreak, *Satyrium behrii* (W.H. Edwards)
- [ ] Sooty Hairstreak, *Satyrium fuliginosa* (W.H. Edwards)
- [ ] California Hairstreak, *Satyrium californica* (W.H. Edwards)
- [ ] Sylvan Hairstreak, *Satyrium sylvinus* (Boisduval)
- [ ] Striped Hairstreak, *Satyrium liparops* (LeConte)
- [ ] Hedgerow Hairstreak, *Satyrium saepium* (Boisduval)
- [ ] Bramble Hairstreak, *Callophrys affinis* (W.H. Edwards)
- [ ] Sheridan's Hairstreak, *Callophrys sheridanii* (W.H. Edwards)
- [ ] Thicket Hairstreak, *Callophrys spinetorum* (Hewitson)
- [ ] Johnson's Hairstreak, *Callophrys johnsoni* (Skinner)
- [ ] Cedar Hairstreak, *Callophrys nelsoni* (Boisduval)
- [ ] Juniper Hairstreak, *Callophrys gryneus* (Hübner)
- [ ] Brown Elfin, *Callophrys augustinus* (Westwood)
- [ ] Western Elfin, *Callophrys iroides* (Boisduval)
- [ ] Moss's Elfin, *Callophrys mossii* (Hy. Edwards)
- [ ] Hoary Elfin, *Callophrys polios* (Cook & Watson)
- [ ] Eastern Pine Elfin, *Callophrys niphon* (Hübner)
- [ ] Western Pine Elfin, *Callophrys eryphon* (Boisduval)
- [ ] Grey Hairstreak, *Strymon melinus* Hübner

Behr's Hairstreak

Blues (Subfamily Polyommatinae)
- [ ] Eastern Tailed Blue, *Cupido comyntas* (Godart)
- [ ] Western Tailed Blue, *Cupido amyntula* (Boisduval)
- [ ] Spring Azure, *Celastrina lucia* (W. Kirby)
- [ ] Western Spring Azure, *Celastrina echo* (W.H. Edwards)
- [ ] Square-spotted Blue, *Euphilotes battoides* (Behr)
- [ ] Arrowhead Blue, *Glaucopsyche piasus* (Boisduval)
- [ ] Silvery Blue, *Glaucopsyche lygdamus* (Doubleday)
- [ ] Northern Blue, *Plebejus idas* (Linnaeus)
- [ ] Anna's Blue, *Plebejus anna* (W.H. Edwards)
- [ ] Melissa's Blue, *Plebejus melissa* (W.H. Edwards)
- [ ] Greenish Blue, *Plebejus saepiolus* (Boisduval)
- [ ] Boisduval's Blue, *Plebejus icarioides* (Boisduval)
- [ ] Acmon Blue, *Plebejus acmon* (Westwood)
- [ ] Cranberry Blue, *Plebejus optilete* (Knoch)
- [ ] Arctic Blue, *Plebejus glandon* (de Pruner)

METALMARKS (Family Riodinidae)
- [ ] Mormon Metalmark, *Apodemia mormo* (C. & R. Felder)

Mormon Metalmark

BRUSH-FOOTED BUTTERFLIES (Family Nymphalidae)
Fritillaries (Subfamily Heliconiinae)
- [ ] Variegated Fritillary, *Euptoeita claudia* (Cramer)
- [ ] Great Spangled Fritillary, *Speyeria cybele* (Fabricius)
- [ ] Callippe Fritillary, *Speyeria callippe* (Boisduval)
- [ ] Zerene Fritillary, *Speyeria zerene* (Boisduval)
- [ ] Aphrodite Fritillary, *Speyeria aphrodite* (Fabricius)
- [ ] Atlantis Fritillary, *Speyeria atlantis* (W.H. Edwards)
- [ ] Northwestern Fritillary, *Speyeria hesperis* (W.H. Edwards)
- [ ] Hydaspe Fritillary, *Speyeria hydaspe* (Boisduval)
- [ ] Mormon Fritillary, *Speyeria mormonia* (Boisduval)
- [ ] Mountain Fritillary, *Boloria alaskensis* (Holland)
- [ ] Bog Fritillary, *Boloria eunomia* (Esper)
- [ ] Silver-bordered Fritillary, *Boloria selene* (Denis & Schiffermüller)
- [ ] Meadow Fritillary, *Boloria bellona* (Fabricius)
- [ ] Frigga Fritillary, *Boloria frigga* (Thunberg)
- [ ] Dingy Fritillary, *Boloria improba* (Butler)
- [ ] Western Meadow Fritillary, *Boloria epithore* (W.H. Edwards)
- [ ] Polar Fritillary, *Boloria polaris* (Boisduval)
- [ ] Alberta Fritillary, *Boloria alberta* (W.H. Edwards)
- [ ] Freija Fritillary, *Boloria freija* (Thunberg)
- [ ] Beringian Fritillary, *Boloria natazhati* (Gibson)
- [ ] Astarte Fritillary, *Boloria astarte* (Doubleday)
- [ ] Distinct Fritillary, *Boloria distincta* (Gibson)
- [ ] Titania Fritillary, *Boloria titania* (Esper)

Crescents and Checkerspots (Subfamily Melitaeinae)

- Northern Crescent, *Phyciodes cocyta* (Cramer)
- Tawny Crescent, *Phyciodes batesii* (Reakirt)
- Field Crescent, *Phyciodes pulchella* (Boisduval)
- Pale Crescent, *Phyciodes pallida* (W.H. Edwards)
- Mylitta Crescent, *Phyciodes mylitta* (W.H. Edwards)
- Northern Checkerspot, *Chlosyne palla* (Boisduval)
- Rockslide Checkerspot, *Chlosyne whitneyi* (Behr)
- Hoffmann's Checkerspot *Chlosyne hoffmanni* (Behr)
- Gillett's Checkerspot, *Euphydryas gillettii* (Barnes)
- Edith's Checkerspot, *Euphydryas editha* (Boisduval)
- Variable Checkerspot, *Euphydryas chalcedona* (Doubleday)

Nymphs (Subfamily Nymphalinae)
Anglewings (Tribe Nymphalini)

- Satyr Comma, *Polygonia satyrus* (W.H. Edwards)
- Green Comma, *Polygonia faunus* (W.H. Edwards)
- Hoary Comma, *Polygonia gracilis* (Grote & Robinson)
- Oreas Comma, *Polygonia oreas* (W.H. Edwards)
- Gray Comma, *Polygonia progne* (Cramer)
- Compton Tortoiseshell, *Nymphalis vaualbum* (Denis & Schiffermüller)
- California Tortoiseshell, *Nymphalis californica* (Boisduval)
- Mourning Cloak, *Nymphalis antiopa* (Linnaeus)
- Milbert's Tortoiseshell, *Aglais milberti* (Godart)
- Painted Lady, *Vanessa cardui* (Linnaeus)
- West Coast Lady, *Vanessa annabella* (Field)
- American Lady, *Vanessa virginiensis* (Drury)
- Red Admirable, *Vanessa atalanta* (Linnaeus)

Admirals (Subfamily Limenitidinae)

- Lorquin's Admiral, *Limenitis lorquini* Boisduval
- White Admiral, *Limenitis arthemis* (Drury)

Satyrs (Subfamily Satyrinae)

- Ringlet, *Coenonympha tullia* Müller
- Common Wood Nymph, *Cercyonia pegala* (Fabricius)
- Great Basin Wood Nymph, *Cercyonis sthenele* (Boisduval)
- Dark Wood Nymph, *Cercyonis oetus* (Boisduval)

Alpines (Genus *Erebia*)

- Common Alpine, *Erebia epipsodea* Butler
- Taiga Alpine, *Erebia mancinus* Doubleday
- Vidler's Alpine, *Erebia vidleri* Elwes
- Ross's Alpine, *Erebia rossii* (Curtis)
- Red-disked Alpine, *Erebia discoidalis* (W. Kirby)
- Magdalena Alpine, *Erebia magdalena* Strecker
- Mt. McKinley Alpine, *Erebia mackinleyensis* Gunder
- Mountain Alpine, *Erebia pawlowskii* Ménétriés

California Tortoiseshell

Arctics (Genus *Oeneis*)
- ☐ Great Arctic, *Oeneis nevadensis* (C. & R. Felder)
- ☐ Macoun's Arctic, *Oeneis macounii* (W.H. Edwards)
- ☐ Chryxus Arctic, *Oeneis chryxus* (Doubleday)
- ☐ Uhler's Arctic, *Oeneis uhleri* (Reakirt)
- ☐ Alberta Arctic, *Oeneis alberta* Elwes
- ☐ Jutta Arctic, *Oeneis jutta* (Hübner)
- ☐ White-veined Arctic, *Oeneis bore* (Schneider)
- ☐ Melissa Arctic, *Oeneis melissa* (Fabricius)
- ☐ Polyxenes Arctic, *Oeneis polyxenes* (Fabricius)
- ☐ Philip's Arctic, *Oeneis philipi* Troubridge

Milkweed Butterflies (Subfamily Danainae)
- ☐ Monarch, *Danaus plexippus* (Linnaeus)

SKIPPERS (Family Hesperiidae)
Spread-winged Skippers (Subfamily Pyrginae)
- ☐ Silver-spotted Skipper, *Epargyreus clarus* (Cramer)
- ☐ Northern Cloudywing, *Thorybes pylades* (Scudder)
- ☐ Dreamy Duskywing, *Erynnis icelus* (Scudder & Burgess)
- ☐ Propertius Duskywing, *Erynnis propertius* (Scudder & Burgess)
- ☐ Pacuvius Duskywing, *Erynnis pacuvius* (Lintner)
- ☐ Afranius Duskywing, *Erynnis afranius* (Lintner)
- ☐ Persius Duskywing, *Erynnis persius* (Scudder)
- ☐ Grizzled Skipper, *Pyrgus centaureae* (Rambur)
- ☐ Two-banded Checkered Skipper, *Pyrgus ruralis* (Boisduval)
- ☐ Common Checkered Skipper, *Pyrgus communis* (Grote)
- ☐ Common Sootywing, *Pholisora catullus* (Fabricius)

Intermediate Skippers (Subfamily Heteropterinae)
- ☐ Arctic Skipper, *Carterocephalus palaemon* (Pallas)

Grass Skippers (Subfamily Hesperiinae)
- ☐ Garita Skipper, *Oarisma garita* (Reakirt)
- ☐ European Skipper, *Thymelicus lineola* (Oshsenheimer)
- ☐ Common Branded Skipper, *Hesperia comma* (Linnaeus)
- ☐ Plains Skipper, *Hesperia assiniboia* (Lyman)
- ☐ Western Branded Skipper, *Hesperia colorado* (Scudder)
- ☐ Juba Skipper, *Hesperia juba* (Scudder)
- ☐ Nevada Skipper, *Hesperia nevada* (Scudder)
- ☐ Sachem Skipper, *Atalopedes campestris* (Boisduval)
- ☐ Peck's Skipper, *Polites peckius* (W. Kirby)
- ☐ Sandhill Skipper, *Polites sabuleti* (Boisduval)
- ☐ Draco Skipper, *Polites draco* (W.H. Edwards)
- ☐ Tawny-edged Skipper, *Polites themistocles* (Latreille)
- ☐ Long Dash, *Polites mystic* (W.H. Edwards)
- ☐ Sonoran Skipper, *Polites sonora* (Scudder)
- ☐ Woodland Skipper, *Ochlodes sylvanoides* (Boisduval)
- ☐ Dun Skipper, *Euphyes vestris* (Boisduval)
- ☐ Common Roadside Skipper, *Amblyscirtes vialis* (W.H. Edwards)

Common Checkered Skipper

Index of Scientific Names

This index lists primary species names and alternate names only.
Where more than one page number is listed, numbers in **boldface** are main entries.

Aglais milberti, 268
Aglais californica.
　See *Nymphalis californica*
Agriades aquilo.
　See *Plebejus glandon*
Agriades glandon.
　See *Plebejus glandon*
Amblyscirtes vialis, 343
Anthocharis, 61
Anthocharis
　sara, 83
　stella, 85
Apodemia mormo, 186
Atalopedes campestris, 334

Basilarchia arthemis.
　See *Limenitis arthemis*
Basilarchia lorquini.
　See *Limenitis lorquini*
Boloria
　alaskensis, 208
　alberta, 222
　astarte, 226
　astarte distincta. See *distincta*
　bellona, 214
　distincta, 228
　epithore, 220
　eunomia, 210
　freija, 224
　frigga, 216
　improba, 218
　napaea. See *alaskensis*
　natazhati, 225
　polaris, 221
　selene, 212
　titania, 230
　toddi. See *bellona*

Callophrys, 143
Callophrys
　affinis, 139
　augustinus, 145
　dumetorum. See *affinis*
　eryphon, 151
　fotis mossii. See *mossii*
　grynea. See *nelsoni, gryneus*
　grynea siva. See *gryneus*
　gryneus, 143, **144**
　iroides, 147
　johnsoni, 142

　mossii, 148
　nelsoni, 143
　niphon, 150
　perplexa affinis. See *affinis*
　polios, 149
　rosneri. See *nelsoni*
　sheridanii, 140
　siva. See *nelsoni, gryneus*
　spinetorum, 141
Callophrys polia.
　See *Callophrys polios*
Carterocephalus palaemon, 326
Celastrina
　argiolus. See *lucia*
　argiolus echo. See *echo*
　argiolus nigrescens. See *echo*
　echo, 160
　ladon. See *lucia*
　ladon echo. See *echo*
　ladon nigrescens. See *echo*
　lucia, 158
　quesnellii. See *lucia*
Cercyonis
　pegala, 279
　oetus, 283
　silvestris. See *sthenele*
　sthenele, 281
Chalceria dione.
　See *Lycaena dione*
Chalceria heteronea.
　See *Lycaena heteronea*
Charidryas damoetus.
　See *Chlosyne whitneyi*
Charidryas hoffmanni.
　See *Chlosyne hoffmanni*
Charidryas palla.
　See *Chlosyne palla*
Charidryas whitneyi.
　See *Chlosyne whitneyi*
Chlosyne, 231, 242
Chlosyne
　hoffmanni, 245
　palla, 242
　whitneyi, 244
Clossiana alberta.
　See *Boloria alberta*
Clossiana bellona.
　See *Boloria bellona*
Clossiana chariclea.
　See *Boloria titania*

Clossiana epithore.
　See *Boloria epithore*
Clossiana eunomia.
　See *Boloria eunomia*
Clossiana freija.
　See *Boloria freija*
Clossiana frigga.
　See *Boloria frigga*
Clossiana improba.
　See *Boloria improba*
Clossiana polaris.
　See *Bolaria polaris*
Clossiana selene.
　See *Boloria selene*
Clossiana titania.
　See *Boloria titania*
Clossiana tritonia.
　See *Boloria astarte*
Coenonympha
　ampelos. See *tullia*
　californica. See *tullia*
　inornata. See *tullia*
　ochracea. See *tullia*
　tullia, 278
Coliadinae (subfamily), 87–88
Colias (genus), 87
Colias
　alexandra, 99
　canadensis, 105
　chippewa. See *palaeno*
　christina, 100
　eurytheme, 91
　gigantea, 93
　hecla, 104
　interior, 94
　meadii, 102
　nastes, 106
　occidentalis, 97
　palaeno, 96
　pelidne, 95
　philodice, 89
Cupido
　amyntula, 156
　comyntas, 154
Cynthia cardui.
　See *Vanessa cardui*

Danainae (subfamily), 307
Danaus plexippus, 56, **308**

INDEX OF SCIENTIFIC NAMES

Epargyreus clarus, 313
Epidemia dorcas.
   See *Lycaena dorcas*
Epidemia helloides.
   See *Lycaena helloides*
Epidemia nivalis.
   See *Lycaena nivalis*
Epidemia mariposa.
   See *Lycaena mariposa*
Erebia (genus), 285–286
Erebia
   disa mancinus. See *mancinus*
   discoidalis, 292
   epipsodea, 287
   mackinleyensis, 294
   magdalena, 293
   mancinus, 289
   pawlowskii, 295
   rossi. See *rossii*
   rossii, 291
   theano. See *pawlowskii*
   vidleri, 290
Erynnis
   afranius, 318
   icelus, 315
   pacuvius, 317
   persius, 319
   propertius, 316
Euchloe, 61
Euchloe
   ausonia. See *ausonides*
   ausonides, 77
   creusa, 79
   hyantis lotta. See *lotta*
   lotta, 82
   naina, 81
Euphilotes battoides, **162**, 133
Euphydryas, 231
Euphydryas
   anicia. See *chaldedona*
   chalcedona, 250
   editha, 248
   gillettii, 246
Euphyes
   ruricola. See *vestris*
   vestris, 342
Euptoeita claudia, 192
Everes amyntula.
   See *Cupido amyntula*
Everes comyntas.
   See *Cupido comyntas*

Glaucopsyche
   lygdamus, 166
   piasus, 164

Harkenclenus titus.
   See *Satyrium titus*

Heliconiinae (subfamily),
   190–191
Hesperia, 329, 332, 333, 334
Hesperia
   assiniboia, 330
   colorado, 331
   comma, 329
   comma assiniboia.
      See *assiniboia*
   comma colorado.
      See *colorado*
   comma harpalus.
      See *colorado*
   comma oregonia.
      See *colorado*
   juba, 332
   nevada, 333
Hesperiidae (family), 310–311
Hesperiinae & Heteropterinae
   (subfamilies.), 325
Hypodryas gillettii.
   See *Euphydryas gillettii*
Hyllolycaena hyllus.
   See *Lycaena hyllus*

Icaricia acmon.
   See *Plebejus acmon*
Icaricia icarioides.
   See *Plebejus icarioides*
Incisalia augustinus.
   See *Callophrys augustinus*
Incisalia augustus.
   See *Callophrys augustinus*
Incisalia eryphon.
   See *Callophrys eryphon*
Incisalia iroides.
   See *Callophrys iroides*
Incisalia mossii.
   See *Callophrys mossii*
Incisalia niphon.
   See *Callophrys niphon*
Incisalia polia.
   See *Callophrys mossii*

Limenitidinae (subfamily),
   274
Limenitis
   arthemis, 273, **276**
   lorquini, 275
Loranthomitoura johnsoni.
   See *Callophrys johnsoni*
Loranthomitoura spinetorum.
   See *Callophrys spinetorum*
Lycaeides anna.
   See *Plebejus anna*
Lycaeides idas. See *Plebejus idas*
Lycaeides idas anna.
   See *Plebejus anna*

Lycaeides melissa.
   See *Plebejus melissa*
Lycaena
   cuprea. See *cupreus*
   cupreus, 114
   dione, 118
   dorcas, 122
   helloides, 124
   heteronea, 120
   hyllus, 116
   mariposa, 128
   nivalis, 126
   phlaeas, 112
   snowi. See *cupreus*
   thoe. See *hyllus*
Lycaenidae (family), 108–109
Lycaeninae (subfamily), 110–111

Melitaea palla.
   See *Chlosyne palla*.
Melitaeinae (subfamily), **231**, 242
Mitoura barryi.
   See *Callophrys gryneus*
Mitoura johnsoni.
   See *Callophrys johnsoni*
Mitoura gryneus.
   See *Callophrys nelsoni*
Mitoura nelsoni.
   See *Callophrys nelsoni*
Mitoura rosneri.
   See *Callophrys nelsoni*
Mitoura siva.
   See *Callophrys nelsoni*
Mitoura spinetorum.
   See *Callophrys spinetorum*

Neophasia menapia, 62
Nymphalidae (family), 188–189
Nymphalinae (subfamily), 252
Nymphalini (tribe), 253
Nymphalis
   antiopa, 266
   californica, 264
   j-album. See *vaualbum*
   l-album. See *vaualbum*
   milberti. See *Aglais milberti*
   vaualbum, 262
   vau-aulbum. See *vaualbum*

Oarisma garita, 327
Ochlodes sylvanoides, 341
Oeneis (genus), 296
Oeneis
   alberta, 301
   bore, 303
   chryxus, 299
   jutta, 302
   macounii, 298

Index of Scientific Names

melissa, 304
nevadensis, 297
philipi, 306
polyxenes, 305
rosovi. See philipi
taygete. See bore
uhleri, 300

Papilio, 40, 42, 49, 57
Papilio
  bairdii. See machaon
  canadensis, 52
  eurymedon, 54
  glaucus canadensis.
    See canadensis
  indra, 48
  machaon, 42, **44**, 46
  multicaudata, 56
  multicaudatus.
    See multicaudata
  oregonius. See machaon
  pikei. See machaon
  rutulus, **50**, 57
  zelicaon, 46
Papilionidae, 32
Papilioninae, 40
Parnassianiinae, 33
Parnassius
  clodius, 36
  eversmanni, 38
  phoebus smintheus.
    See smintheus
  smintheus, 34
Phaedrotes piasus.
  See Glaucopsyche piasus
Pholisora catullus, 324
Phyciodes, 231
Phyciodes
  barnesi. See pallida
  batesii, 234
  campestris. See puchella
  cocyta, 232
  montana. See pulchella
  morpheus. See cocyta
  mylitta, 240
  pallida, 238
  pallidus. See pallida
  pascoensis. See cocyta
  pratensis. See pulchella
  pulchella, 236
  pulchellus. See pulchella
  selenis. See cocyta
Pieridae (family), 58–59
Pierinae (subfamily), 60–61
Pieris
  angelika, 74
  marginalis, 73
  napi. See oleracea, marginalis

oleracea, 72
rapae, 75
Plebeius acmon.
  See Plebejus acmon
Plebeius acquilo.
  See Plebejus glandon
Plebeius argyrognomon.
  See Plebejus idas
Plebeius glandon.
  See Plebejus glandon
Plebeius icarioides.
  See Plebejus icarioides
Plebeius idas. See Plebejus idas
Plebeius melissa.
  See Plebejus melissa
Plebeius optilete.
  See Plebejus optilete
Plebeius saepiolus.
  See Plebejus saepiolus
Plebejus
  acmon, 178
  argyrognomon. See idas
  anna, 170
  glandon, 182
  icarioides, 176
  idas, 168
  melissa, 172
  optilete, 180
  saepiolus, 174
Polites, 329, 338
Polites
  coras. See peckius
  draco, 337
  mystic, 339
  peckius, 335
  sabuleti, 336
  sonora, 340
  themistocles, 338
Polygonia, 253
Polygonia
  faunus, 256
  gracilis, 257
  hylus. See faunus
  oreas, 259
  progne, 259, **260**
  satyrus, 254
  sylvius. See faunus
  zephyrus. See gracilis
Polyommatinae (subfamily), 153
Pontia, 61
Pontia
  beckerii, 68
  occidentalis, 64
  protodice, 66
  sisymbrii, 70
Proclossiana eunomia.
  See Boloria eunomia

Pyrginae (subfamily), 312
Pyrgus, 320, 321, 323
Pyrgus
  centaureae, 320
  communis, 322
  ruralis, 321

Riodinidae (family), 184–185
Roddia l album.
  See Nymphalis vaualbum

Satyrinae (subfamily), 277
Satyrium
  behrii, 133
  californica, 135
  californicum. See californica
  fuliginosa, 134
  fuliginosum. See fuliginosa
  liparops, 137
  saepium, 138
  sylvinum. See sylvinus
  sylvinus, 136
  titus, 132
Speyeria
  aphrodite, 200
  atlantis, 202. See also hesperis
  callippe, 196
  cybele, 194
  electa. See hesperis
  hesperis, 204
  hydaspe, 206
  leto. See cybele
  mormonia, 207
  zerene, 198
Strymon liparops.
  See Satyrium liparops
Strymon melinus, 152

Theclinae (subfamily), 130–131
Thorybes pylades, 314
Thymelicus lineola, 328

Vacciniina optilete.
  See Plebejus optilete
Vanessa, 253
Vanessa
  annabella, 271
  atalanta, 273
  cardui, 269
  carye. See annabella
  virginiensis, 272

Index of Common Names

This index lists primary species names and alternate names only, by group.
Where more than one page number is listed, numbers in **boldface** are main entries.

Admiral
  Lorquin's, 275
  Orangetip. *See* Lorquin's
  Red-spotted. *See* White
  Red-spotted Purple.
    See White
  White, 273, **276**
Admirals, 274
Alpine
  Banded. *See* Mountain
  Butler's. *See* Common
  Cascades. *See* Vidler's
  Common, **287**, 290, 291
  Magdalena, 286, **293**, 294
  Mancina. *See* Taiga
  Mountain, 295
  Mt. McKinley, 294
  Northwest. *See* Vidler's
  Red-disked, 287, **293**
  Rockslide. *See* Magdalena
  Ross'. *See* Ross's
  Ross's, 291
  Taiga, 289
  Theano. *See* Mountain
  Two-dot. *See* Ross's
  Vidler's, **290**, 291, 297
  Yellow-dotted. *See* Mountain
Alpines, 285–286, 296
Anglewing. *See* Comma
**Anglewings & their Relatives,
    253.** *See also* Comma; Lady;
    Nymphs; Tortoiseshell
Arctic
  Alberta, 301
  Artic Grayling.
    See White-veined
  Baltic Grayling. *See* Jutta
  Banded. *See* Polyxenes
  Brown. *See* Chryxus
  Canada's. *See* Macoun's
  Chryxus, **299**, 300
  Early. *See* Philip's
  Felder's. *See* Great
  Forest. *See* Jutta
  Giant. *See* Great
  Great, **297**, 298
  Great Grayling. *See* Great.
  Jutta, 289, **302**
  Macoun's, 298

Melissa, **304**, 305
Mottled. *See* Melissa
Nevada. *See* Great
Pacific. *See* Great
Philip's, 306
Polyxenes, 304, **305**
Prairie. *See* Alberta
Rocky Mountain. *See* Uhler's
Rosov's. *See* Philip's
Uhler's, **300**, 301
White-veined, **303**, 304, 305
Arctics, 296
Azure
  Boreal Spring. *See* Spring
  Spring, 146, 153, **158**, 160,
    161, 224
  Western Spring, 160

Blue
  Acmon, 162, **178**
  Anna. *See* Anna's
  Anna's, 168, 169, **170**, 173,
    174
  Arctic, 182
  Arrowhead, 164
  Boisduval's, 174, **176**, 179
  Cranberry, 180
  Eastern Tailed, **154**, 156
  Greenish, 89, **174**
  Icarioides. *See* Boisduval's
  Melissa. *See* Melissa's
  Melissa's, 162, 168, **172**
  Northern, 162, **168**, 170,
    171, 172, 178
  Rustic. *See* Arctic
  Silvery, 165, **166**
  Square-spotted, 133, **162**
  Western Tailed, 154, **156**
  Yukon. *See* Cranberry
Blues, 153. *See also* Azure
**Brush-footed Butterflies,
    188–189.** *See also* Admiral;
  Alpine; Anglewings; Arctic;
  Comma; Checkerspot;
  Crescent; Duskywing;
  Fritillary; Lady; Milkweed
  Butterflies; Monarch;
  Mourning Cloak; Nymphs;
  Satyrs; Tortoiseshell

Checkered Skipper
  Alpine. *See* Grizzled
  Common, 320, **322**
  Two-banded, 321
  Common Sootywing, 324
Checkerspot
  Alpine. *See* Rockslide
  Anicia. *See* Variable
  Aster. *See* Hoffmann's
  Chalcedon. *See* Variable
  Creamy. *See* Northern
  Damoetus. *See* Rockslide
  Edith's, 247, **248**, 250
  Gillette's. *See* Gillett's
  Gillett's, **246**, 250
  Hoffmann's, 242, **245**
  Northern, 242
  Pacific. *See* Hoffmann's
  Pale. *See* Northern
  Ridge. *See* Edith's
  Rockslide, 242, **244**
  Variable, 247, **250**
  Whitney's. *See* Rockslide
  Yellowstone. *See* Gillett's
Comma
  Dark Gray Anglewing.
    See Oreas
  Faun. *See* Green
  Faun Anglewing.
    See Green
  Faunus Anglewing.
    See Green
  Golden Anglewing.
    See Satyr
  Gray, 259, **260**
  Green, 256
  Green Anglewing.
    See Green
  Grey. *See* Gray
  Hoary, **257**, 259
  Hoary Anglewing.
    See Hoary
  Hope Butterfly. *See* Satyr
  Oreas, **259**, 261
  Oreas Anglewing.
    See Oreas
  Oreas Angle-wing.
    See Oreas
  Satyr, 254

Satyr Anglewing. *See* Satyr
Silenus Anglewing.
  See Orcas
The Satyr. *See* Satyr
Zephyr. *See* Hoary
Zephyr Anglewing.
  See Hoary
Common Sootywing.
  See Checkered Skipper
Copper
  American, **112**, 114, 129
  Blue, **120**, 134
  Bronze, 116
  Dione. *See* Gray
  Dorcas, **122**, 124, 128
  Gray, 118
  Lustrous, 114
  Lilac-bordered, **126**, 128
  Mariposa, 128
  Purplish, 122, 123, **124**, 128
  Small. *See* America
Coppers, 110–111
Crescent
  Field, **236**, 238
  Field Crescentspot.
    See Field
  Mylitta, 238, 239, **240**
  Northern, **242**, 236
  Northern Pearl. *See* Northern
  Orange. *See* Northern
  Pale, **238**, 240
  Pale Crescentspot. *See* Pale
  Pasco. *See* Northern
  Pearl. *See* Northern
  Pearly Crescentspot.
    See Northern
  Tawny, **234**, 236, 238, 239
  Thistle. *See* Mylitta
Crescents & Checkerspots, 231

Duskywing
  Afranius, 317, **318**, 319
  Aspen. *See* Dreamy
  Bald. *See* Afranius
  Buckthorn. *See* Pacuvius
  Dreamy, 315
  Dyar's. *See* Pacuvius
  Hairy. *See* Persius
  Pacuvius, **317**, 319
  Persius, 317, 318, **319**
  Propertius, 316
  Western Oak. *See* Propertius

Elfin
  Brown, **145**, 147
  Eastern Pine, **150**, 151
  Hoary, **149**, 151

Moss's, 148
Western, 147
Western Pine, 150, **151**

Field Crescentspot.
  See Crescent
Fritillary
  Alberta, **222**, 226
  Albert's. *See* Alberta
  Aphrodite, **200**, 202, 205
  Arctic. *See* Titania
  Astarte, **226**, 229
  Atlantis, 199, **202**, 204, 205
  Beringian, 225
  Boeber's. *See* Astarte
  Bog, 209, **210**, 217
  Callippe, **196**, 198
  Cryptic. *See* Beringian
  Dingy, **218**, 220
  Distinct, 227, **228**
  Electa. *See* Northwestern
  Freija, 216, 217, **224**, 225
  Frigga, **216**, 218, 220
  Great Spangled, **194**, 202
  Hydaspe, 206
  Lavender. *See* Hydaspe
  Leto. *See* Great Spangled
  Meadow, **214**, 220
  Mormon, 207
  Mountain, 208
  Northwestern, 200, 202, 203, **204**
  Northwestern Silverspot.
    See Northwestern
  Pacific. *See* Western Meadow
  Polar, 221
  Polaris. *See* Polar
  Purplish Lesser.
    See Titania
  Silver-bordered, 212
  Titania, 230
  Variegated, 190, **192**, 213
  Western Meadow, 215, **220**
  Zerene, 198
Fritillaries, 190–191

Gossamer-winged Butterflies, 108–109. *See also* Azure; Blue; Copper; Elfin; Hairstreak
Grass Skippers, 325

Hairstreak,
  Barry's Juniper. *See* Juniper
  Behr's, **133**, 136
  Bramble, **139**, 140
  California, **135**, 136

Cedar, 139, **143**, 144
Coral, **132**, 136
Gray. *See* Grey
Grey, 152
Hedgerow, 136, **138**
Immaculate Bramble.
  See Bramble
Immaculate Green.
  See Bramble
Johnson's, 139, **142**
Juniper, 139, 141, 143, **144**
Nelson's Juniper.
  See Cedar
Rosner's. *See* Cedar
Rosner's Juniper.
  See Cedar
Sheridan's, 140
Siva. *See* Juniper
Siva Juniper. *See* Juniper
Sooty, **134**, 136
Sooty Gossamer Wing.
  See Sooty
Striped, 136, **137**
Sylvan, 135, **136**
Thicket, 139, **141**, 143
Western Green.
  See Bramble
White-lined Sheridan's.
  See Sheridan's
Hairstreaks & Elfins, 130–131.
  See also Elfins; Hairstreaks

Lady
  Alderman. *See* Red Amirable
  American, 272
  American Painted.
    See American
  Cosmopolite. *See* Painted
  Cynthia. *See* Painted
  Cynthia of the Thistle.
    See Painted
  Hunter's Butterfly.
    See American
  Nettle Butterfly. *See* Red Admirable
  Painted, 253, **269**, 271, 272
  Red Admirable, 253, **273**, 274, 284, 302
  Red Admiral.
    See Red Admirable
  Thistle Butterfly.
    See Painted
  Virginia. *See* American
  West Coast, **271**, 272, 273
  Western Painted.
    See West Coast
Long Dash. *See* Skipper

Marble
  Dappled. *See* Large
    Desert, 82
  Green, 81
  Large, **77**, 79, 82
  Marblewing, Creamy.
    See Large
    Northern, 79
  Western. *See* Desert
Metalmark, Mormon, 186
Metalmarks, 184–185
Milkweed Butterflies, **307**
Monarch, 56, 307, **308**
Mourning Cloak.
  See Tortoiseshell

Northern Cloudywing.
  See Skipper
Nymphs, 252. *See also*
  Anglewings; Comma;
  Lady; Tortoiseshell

**Old World Swallowtail Group,
  42–43**, 47, 48, 49.
  See also Swallowtail;
  Tiger Swallowtails
Orangetip
  Sara, **83**, 85, 86
  Stella, 83, 84, **85**

Pale Crescentspot. *See* Crescent
Parnassian
  Clodius, 36
  Eversmann's, 36, **38**
  Pheobus. *See* Rocky
    Mountain
  Rocky Mountain, 34
Parnassians, 32, **33**, 40
Pearly Crescentspot.
  See Crescent
Ringlet. *See* Satyrs

Satyrs, 277
  California Ringlet.
    See Ringlet
  Common Ringlet.
    See Ringlet
  Common Wood Nymph,
    279, 281
  Dark Wood Nymph, 283
  Goggle-eyed Wood Nymph.
    See Common Wood
    Nymph
  Great Basin Wood Nymph,
    281, 283, 284
  Inornate Ringlet.
    See Ringlet

Large Heath. *See* Ringlet
Large Wood Nymph. *See*
  Common Wood Nymph
Ochre Ringlet. *See* Ringlet
Ox-eyed Wood Nymph. *See*
  Common Wood Nymph
  Ringlet, 278
Scrub Wood Nymph. *See*
  Great Basin Wood Nymph
Small Wood Nymph. *See*
  Great Basin Wood Nymph,
  Dark Wood Nymph
Woodland Satyr. *See* Great
  Basin Wood Nymph
Skipper
  Arctic, 325, **326**
  Arctic Skipperling.
    See Arctic
  Black Little. *See* Common
    Roadside
  Checkered. *See* Common
    Checkered Skipper
  Chequered. *See* Arctic
  Colorado. *See* Western
    Branded
  Common Branded, **329**,
    330, 331
  Common Roadside, 343
  Draco, 337
  Dragon. *See* Draco
  Dun, 342
  Essex. *See* European
  European, 327, **328**, 341
  European Skipperling.
    See European
  Field. *See* Sachem
  Garita, 327
  Garita Skipperling.
    See Garita
  Grizzled Skipper, 320
  Holarctic Grass.
    See Common Branded
  Jagged-bordered. *See* Juba
  Juba, 332
  Long Dash, **339**, 341
  Montane. *See* Nevada
  Mystic. *See* Long Dash
  Nevada, 333
  Northern Cloudywing, 314
  Peck's, 335
  Plains, 329, **330**
  Roadside.
    See Common Roadside
  Roadside Rambler.
    See Common Sootywing
  Rocky Mountain.
    See Draco
  Sachem, 334

Saltgrass. *See* Sandhill
Sandhill, 336
Silver-spotted, 312, **313**,
  315
Sonoran, 340
Sun Sedge. *See* Dun.
Tawny-edged, 338
Western. *See* Woodland
Western Branded, 329,
  330, **331**
Western Long Dash.
  See Sonoran
Western Skipperling.
  See Garita
Woodland, 339, **341**
Yellowpatch. *See* Peck's
Yuba. *See* Juba
Skippers, 310–311. *See also*
  Duskywing; Grass Skippers;
  Spread-winged Skippers
**Spread-winged Skippers,
  312**
Sulphur
  Alexandra's, 88, 97, **98**,
    100, 101
  Canada, 88, 104, **105**
  Chippewa. *See* Palaeno
  Christina's, 88, 98, 99, **100**
  Clouded, 88, 91, **89**, 94, 97,
    280
  Common. *See* Clouded
  Giant, 88, **93**, 94, 97
  Hecla, 88, **104**, 107
  Mead's, 88, **102**
  Nastes, 87, 88, 95, 96, **106**
  Orange, 88, **91**, 94
  Palaeno, 88, 94, **96**
  Pelidne, 88, 94, **95**
  Pink-edged, 88, 93, **94**,
    95, 97
  Yellow. *See* Clouded
  Western, 88, **97**, 100
Sulphurs, 87
Swallowtail
  Anise, **46**, 51
  Baird. *See* Old World
  Indra, 48
  Old World, **44**, 46, 137
  Oregon. *See* Old World
  Zelicaon. *See* Anise.
Swallowtails, 40–41. *See also*
  Old World Swallowtail
  Group; Tiger Swallowtail;
  Tiger Swallowtail Group

Tiger Swallowtail
  Canadian, 50, 51, **52**, 57
  Pale, 50, **54**

Index of Common Names

Two-tailed, 51, **56**
Western, 49, **50**, 53, 57
Tiger Swallowtail Group, 43, 46, **49**, 56. *See also* Tiger Swallowtail
Tortoiseshell
American. *See* Milbert's
Antiopa. *See* Mourning Cloak
California, 264
Camberwell Beauty. *See* Mourning Cloak
Common. *See* Compton
Compton, 262
Compton Tortoise. *See* Compton
Compton's. *See* Compton
False Comma. *See* Compton
Fire-rim. *See* Milbert's
Grand Surprise. *See* Mourning Cloak
Mourning Cloak, 146, 253, 263, **266**, 268
Milbert's, 268
Nettle. *See* Milbert's
Spiny Elm Caterpillar. *See* Mourning Cloak
Western. *See* California
White Petticoat. *See* Mourning Cloak
Yellow Edge. *See* Mourning Cloak

White
Arctic, 74
Becker's, 68
Cabbage, 61, 67, 72, 73, **75**, 87, 280
Checkered, 64, 65, **66**, 71
Great Basin. *See* Becker's
Margined, 73
Mustard. *See* Veined, Margined
Pine, 61, **62**
Sage. *See* Becker's
Spring, 70
Veined, 72
Western, 64
Whites, Marbles & Orangetips, 60–61; *See also* Marble; Orangetip; White
Whites & Sulphurs, 58–59, 69. *See also* White; Whites, Marbles & Orangetips; Sulphur
Wood Nymph. *See* Satyrs

Zerene Fritillary

About The Authors

John Acorn

John Acorn has been a "critter fanatic" his entire life. He is perhaps best known as the writer and host of the television series *Acorn, The Nature Nut*. These days, John lectures at the University of Alberta, serves as spokesperson for the Royal Tyrrell Museum and travels widely as a public speaker. He still finds time for entomology, and is an Associate of the E.H. Strickland Entomology Museum. John has written 15 books, including many well-received field guides, and is the recipient of numerous awards, including the University of Alberta's Distinguished Alumni Award. He lives in Edmonton, with his wife, Dena, and their two young boys Jesse and Benjamin.

Ian Sheldon

Ian Sheldon is a Canadian-born artist, and works alongside John. He has lived in South Africa, Singapore and England, and has travelled much of the world. Exposure to so many bugs inspired him to study them further, earning him the Prince Philip Prize from the Zoological Society of London, and an honours degree from the University of Cambridge. Following his Master of Science degree at the University of Alberta, he decided to devote his attention to the fine arts as well as illustrating entomological field guides. Ian is a recognized artist represented by galleries internationally, and has many natural history books under his belt as both writer and illustrator, including Lone Pine's *Bugs of British Columbia*.